American Institute of Chemical Engineers

Systems Biology

Topical Conference at the 2007 AIChE Annual Meeting

November 4-9, 2007
Salt Lake City, Utah, USA

Printed from e-media with permission by:

Curran Associates, Inc.
57 Morehouse Lane
Red Hook, NY 12571
www.proceedings.com

ISBN: 978-1-60423-838-9

Some format issues inherent in the e-media version may also appear in this print version.

ISBN: 978-1-60423-838-9

American Institute of Chemical Engineers
Proceedings
Three Park Avenue
New York, NY 10016-5991
Phone: 212-591-8100

www.aiche.org

American Institute of Chemical Engineers

Systems Biology
2007

TABLE OF CONTENTS

383a Combinatorial Biosynthesis Of Targeted Tetracycline Polyketides Through Protein Engineering

Wenjun Zhang and Yi Tang. Chemical and Biomolecular Engineering, University of California at Los Angeles, 420 Westwood Plaza, 5667 Boelter Hall, Los angeles, CA 90095

Combinatorial biosynthesis is an effective way to produce valuable unnatural natural products with improved biological activities. However the vast potential of combinatorial biosynthesis is far from being realized due to the lack of general programming rules for recombining enzymes from different sources. We will develop protein engineering tools to understand these intrinsic limitations and engineer natural product biosynthesis. Three interconnected components in the metabolite-responsive regulation circuitry will be used to allow host cell behavior to be dictated by the presence of a specific metabolite, which is biosynthesized by the engineered pathway. The first component is a reconstituted pathway in an organism that is genetically amendable to the directed evolution techniques. The second component is a regulatory cascade that responds to a target metabolite. The third component of the design is a combinatorial library of genes encoding mutants of the target enzyme. This design will enable powerful selection of desired enzyme that can function with a specific pathway and its product. We will use the tetracycline-TetR pair as the proof-of-concept system to study the robustness of the cascade.

383b Single-Cell Zeroth-Order Ssra Protein Degradation Kinetics Enhances the Robustness of Synthetic Gene-Metabolic Oscillator

Wilson W. Wong, Chemical Engineering, UCLA, 5531 Boelter Hall, Los Angeles, CA 90095, Tony Tsai, Biomathematics, UCLA, 5531 Boelter Hall, Los Angeles, CA 90095, and James C. Liao, UCLA, 5531 Boelter Hall, Los Angeles, CA 90095.

Design of synthetic biological circuits often requires protein degradation to accomplish the reverse function in response to a negative signal. In Escherichia coli, protein degradation in the synthetic circuit is commonly achieved through the ssrA-tagged degradation system. In this work, we show that that the degradation kinetics for the native ssrA tag in each cell is in fact zeroth-order, rather than the commonly assumed first order. When measured in a population, the heterogeneity of the cells distorted the true kinetics and exhibited a first-order protein degradation kinetics as a population average. Furthermore we demonstrate that the zeroth-order kinetics significantly enhances the robustness of the gene-metabolic oscillator designed previously.

383c Towards Microbial Synthesis Of Glucaric Acid

Tae Seok Moon[1], Leah Octavio[2], and Kristala Jones Prather[1]. (1) Chemical Engineering, Massachusetts Institute of Technology, 77 Massachusetts Avenue, Room 66-425, Cambridge, MA 02139, (2) Computational and Systems Biology, Massachusetts Institute of Technology, 77 Massachusetts Avenue, Room 66-425, Cambridge, MA 02139

The growing interest in a "biomass-based" economy has led to new efforts to construct and improve microorganisms capable of producing new chemicals. While the current focus is largely on liquid biofuels, a successful "biorefinery" is likely to be a mixed-product facility, with many compounds produced from a single, biomass-derived feed. One such "value-added" compound is glucaric acid, a molecule with potential uses ranging from human therapeutics to polymer synthesis. Glucaric acid is a natural product, found in fruits, vegetables, and mammals, but there is no known microbial pathway. We are working towards the construction of two different routes for the synthesis of glucaric acid from glucose in E. coli. The first is a three-step pathway consisting of a series of enzyme reactions derived from yeast, mammalian, and bacterial sources. While the combination of these three steps into a single organism constitutes a "novel" pathway, each step is a naturally-occurring reaction, demonstrated in the original species. In contrast, the second pathway proposes putative transformations based on the selection of enzymes primarily according to generalized enzyme reactions and not specific substrate-enzyme pairs. This latter method follows a framework of synthetic biology for "retro-biosynthetic design" of novel pathways and requires that new activities consisting of enzymes with altered substrate specificity be created. We will present progress on the construction of these pathways towards the microbial synthesis of glucaric acid.

383d Tat-Regulated Promoter Occupancy Mediates Open Loop Stochastic Bimodality In HIV1 Gene Expression And Latency

John C. Burnett[1], Ron Skupsky[2], Adam P. Arkin[3], and David V. Schaffer[1]. (1) Chemical Engineering, University of California at Berkeley, Department of Chemical Engineering, 201 Gilman Hall, Berkeley, CA 94720-1462, (2) California Institute for Quantitative Biomedical Research, University of California at Berkeley, Berkeley, CA 94720, (3) Bioengineering, University of California at Berkeley, Berkeley, CA 94720-1462

HIV, the retrovirus that causes AIDS, can establish rare, latent infections of cells, and the resulting latent pools of virus represent the most significant barrier to elimination of virus from a patient since they persist for decades and can reactivate at any time. After HIV enters a cell, it semi-randomly integrates its genetic material into the host genome and then utilizes the transcriptional machinery from the host cell to regulate its gene expression. Furthermore, efficient HIV transcriptional regulation also requires the viral protein Tat, a protein encoded within in the HIV genome. Since the viral protein Tat helps catalyze the transcription of HIV genome, this constitutes a strong positive feedback circuit.

We have conducted a rigorous experimental and computational analysis of the gene expression dynamics in a retroviral model of HIV-1. This model retroviral system is derived from HIV-1 and is missing all viral pathogenic functions but contains its ability to integrate chromosomally, its full enhancer and promoter elements, and its implementation of the Tat-mediated positive feedback loop governing transcriptional activation. We have previously reported that stochastic fluctuations in Tat lead to phenotypic diversity, or stochastic bimodality, in HIV-1 gene expression, a phenomenon that can be observed experimentally and predicted computationally with stochastic modeling. We have hypothesized that Tat-mediated feedback is activated sporadically such that after the virus infects a target activated T cell and integrates into the host genome, significant periods of time may elapse before viral expression reaches the point where reproduction and propagation can occur. This time could be enough to allow an activated T cell to transition to its memory state, thereby trapping the lentivirus in an inactive form until such time as that memory cell is reactivated.

We now build on our previous discovery with experimental and computational approaches which utilize a mechanistic model of the HIV-1 promoter, regulated by cis-regulatory factors such as NFκB and Sp1, in addition to viral transactivator Tat. We show that stochastic bimodality may result from a combination of occupancy states of this promoter. Ultimately, these findings may give insight into the establishment of HIV-1 latency and may lead to the mechanistic understanding and design of therapies to purge and eradicate latent HIV-1 reservoirs.

383e Towards Programming Network Connectivities: Ligand-Regulated Shrna Switches

Chase L. Beisel, Chemical Engineering, California Institute of Technology, Pasadena, CA 91125 and Christina Smolke, Chemistry and Chemical Engineering, California Institute of Technology, 236A Spalding, Mail Code 210-41, Pasadena, CA 91125.

Systems biology relies on perturbations to endogenous genetic networks in order to study the effects of network topology and dynamics on the system response over short and long time scales. These perturbations often involve mutation or deletion of designated nodes (genes), and subsequent characterization of system response is often performed under different environmental conditions. An alternative approach is to introduce new connectivities that would provide unique insights into global network behavior and a means to expand upon the network's existing regulatory capabilities. Additional benefits would be provided if the dynamics of the synthetic connections can be specified and tightly controlled. Research within synthetic biology has produced an assortment of complex regulatory networks that have been integrated into endogenous networks. These synthetic networks function almost entirely at the transcriptional level, which restricts how easily they can be implemented as linkages between any given nodes within an endogenous network. Post-transcriptional control can overcome this restriction, especially with the discovery of effective gene silencing mechanisms such as RNA interference (RNAi). The RNAi pathway has been used to orchestrate the sequence-specific degradation of target transcripts, thereby regulating expression levels of the target genes. Practically any gene can be targeted through this mechanism, suggesting that RNAi may serve as a foundation for introducing network linkages between any given nodes.

To this end we have engineered shRNAs (activators of RNAi) so that their activity is regulated by a specified ligand. The resulting molecules, termed shRNA switches, are comprised of an shRNA domain, an aptamer domain, and a linker domain. shRNA switches can adopt two conformations that differ in their ability to be processed by the RNAi machinery. The active conformation includes a formed shRNA domain and a disrupted aptamer domain, whereas the inactive conformation includes a disrupted shRNA domain and a formed aptamer domain. Ligand binding to the aptamer domain stabilizes the inactive conformation, thereby modulating the relative expression level of the target gene. Ligand responsiveness is mediated by the aptamer domain, which is distinct from the shRNA domain and modular; namely, it can be replaced with different aptameric sequences to enable gene regulation through potentially any biomolecule within the cell. Based on the seemingly limitless number of endogenous targets afforded by aptamers and RNAi, shRNA switches have the potential to serve as linkages between any nodes in an endogenous network.

We will discuss our work investigating the mechanism behind ligand regulation of shRNA switch activity and the potential that these molecules hold as ligand-responsive gene regulatory systems as demonstrated within a model system of mammalian cells. Through our work, we have delineated a set of rules that govern the behavior of each shRNA switch and provide a significant link between sequence and function. A simple mathematical model based upon the described mechanism was also developed that accurately represents experimental data and can serve as a predictive tool. We will show that a combination of rational design and mathematical modeling guides the forward engineering of new shRNA switches in a variety of applications, including intelligent therapeutics and the study of large-scale genetic networks.

383f Fabrication Of Anti-Obesity Stilbene Analogues Using Engineered Microbial Platforms: Structure-Function Properties Towards Inhibition Of Digestive Enzymes

Effendi Leonard[1], Chin Giaw Lim[2], Zacchary Fowler[2], Kok Hong Lim[2], and Mattheos Koffas[2]. (1) Chemical Engineering, Massachusetts Institute of Technology (MIT), 77 Massachusetts Ave, Cambridge, MA 02139, (2) Chemical and Biological Engineering, State University of New York, 303 Furnas Hall, Buffalo, NY 14260

Resveratrol is a type of stilbene polyketite found in small quantities in plants. Recently, resveratrol has been demonstrated to increase the life-span of many organisms and is being tested as a new therapeutic against diseases associated with aging, such as diabetes and obesity. The minute quantities of stilbenes in plants hinder the development of inexpensive pharmaceuticals. In the present work, Escherichia coli was engineered to synthesize plant-specific stilbenes and their derivatives that do not exist in nature. The capacity of synthesizing stilbenes from the recombinant E. coli was initially low. The overcoming of this bottleneck was two folds. First, we designed biosynthetic enzyme analogues to search for chimeras that robustly catalyzed stilbene biosynthesis in the microorganism. Second, to accelerate the in vivo reaction rate, fluxes toward the malonyl-CoA backbone was increased. Introduction of a foreign carbon assimilation pathway increased product synthesis. However, high-level synthesis was only achieved through partial repression of lethal metabolic targets. These targets were mapped by a constructed in silico metabolism and gene deletion routines. Other gene targets that result in improvement of stilbene biosynthetic phenotype were also identified. Overall, through these strategies, the E. coli factory was capable of stilbene production in high-levels. Robust synthesis of the various non-natural molecules also allowed the structure-function analysis of inhibitory properties against digestive enzymes responsible of glucose assimilation. A stilbene derivative that exhibited higher potency to inhibit digestive enzymes than a currently prescribed drug was identified.

383g Engineering The Salmonella Type Three Secretion System For Heterologous Protein Production

Danielle Tullman-Ercek[1], Daniel Widmaier[1], Ethan Mirsky[1], Karsten Temme[2], and Christopher Voigt[1]. (1) Pharmaceutical Chemistry, University of California - San Francisco, 1700 4th St., Byer's Hall Rm 409, MC 2540, San Francisco, CA 94158, (2) Bioengineering, University of California - San Francisco, 1700 4th St., Byer's Hall Rm 409, MC 2540, San Francisco, CA 94158

We have engineered a strain of *Salmonella* that can manufacture heterologous proteins of interest and secrete them outside of the cell. Stable expression of these non-native proteins is achieved by using DNA synthesis to build genes that are simultaneously optimized for genetic stability and expression in eubacteria. The resulting proteins are exported from the cell using the type III secretion system encoded on *Salmonella* Pathogeneity Island 1 (SPI-1), which is naturally utilized for the secretion of virulence proteins. To exploit the SPI-1 secretion process, a synthetic genetic system was constructed that interfaces with native SPI-1 regulation, such that heterologous protein expression only occurs when the secretion machinery is assembled and fully functional. Thus, expression is avoided until the cell is capable of exporting the product. Furthermore, we developed fluorescent reporter fusions to the various promoters of the SPI-1 regulatory circuit. We first verified the natural dynamics of the network as reported in the literature and then used the reporters to study various perturbations to the system, with the ultimate goal of increasing heterologous protein secretion levels. The resulting hybrid natural-synthetic system produces significant yields and a novel route to secreted recombinant proteins of biotechnological interest.

439a From Reaction Engineering to Systems Biology

Bernhard O. Palsson, University of California San Diego, 417 Powell-Focht Bioengineering Hall, 9500 Gilman Drive., La Jolla, CA 92093-0412

Cellular functions rely on complex reaction networks, that typically involve hundreds if not thousands of reactions. Topologically these biochemical reaction networks are different from chemical reaction networks in that they have co-factors that participate in many reactions and thus they are highly interconnected. We will describe the reconstruction of genome-scale metabolic reaction networks, how to represent them mathematically, how to compute their likely flux states, and how such computations can be used to drive biological discovery and design.

439b Do We Know Enough to Reconstruct Life in Silico?

Michael L. Shuler[1], Jordan C. Atlas[1], and Evgeni V. Nikoleav[2]. (1) School of Chemical and Biomolecular Engineering, Cornell University, 120 Olin Hall, Ithaca, NY 14853, (2) Department of Biomedical Engineering, Cornell University, Ithaca, NY 14853

The question of "What is essential for life?" is one of the most fundamental questions humankind faces. If we fully understand a system, we should be able to reconstruct it. We propose to construct *in silico* a model of a minimal chemoheterotrophic bacterial cell. Such a model is key to fully understanding and identifying underlying regulatory and organizational concepts central to life. The success of whole organism genome sequencing and high-throughput measurements provides an opportunity for system-level analysis of whole organisms, or what has been termed "systems biology". Systems biology investigates the behavior of all of the elements in a biological system while it is functioning. As a systems biology approach, the Minimal Cell Model (MCM) depicts the total functionality of a minimal cell and its explicit response to perturbations in its environment. In this paper we describe our approach to creating such a model and results demonstrating feasibility. We also discuss constraints on identifying all essential functions.

I. Purpose We propose a dynamic modeling framework to integrate genomic detail and cellular physiology within functionally complete 'hybrid' bacterial cell models. An initial step in this approach is the development of a whole-cell coarse-grained model which explicitly links DNA replication, metabolism, and cell geometry with the external environment. A hybrid model can then be constructed from chemically-detailed and genome-specific subsystems, called modules, inserted into the original coarse-grained model. We have demonstrated this possibility with modules for nucleotides [1] and lipids [2]. We use the sensitivity analysis of the original coarse-grained model to identify which pseudo-molecular processes should be de-lumped into molecularly detailed mathematical modules to implement a particular biological function. A minimal cell is a hypothetical entity defined by the essential functions required for life [1]. Although others have the goal of experimentally constructing a minimal cell [3], we seek to identify a minimum number of genes necessary and sufficient for the cell to divide and grow continuously in a rich environment with preformed nutrients and constant temperature and pH. The model, which contains kinetic, thermodynamic, and stoichiometric constraints, is used as a tool to identify the organizing principles which relate the dynamic non-linear functioning of the cell to the genome sequence. We recognize that a unique set of minimal genes may not exist, but we believe a unique set of essential functions exist and that we can generate a candidate minimal gene set to accomplish these functions. Functions may be chemical or physical such as constraints on DNA packing.

II. Approach The project proposed here includes three main parts: 1) Development of novel algorithms for stability and sensitivity analysis of hybrid cell models, 2) Implementation of a system for parameter estimation, and 3) Construction of a genomically and chemically detailed Minimal Cell Model.

III. Results Using an existing coarse-grained model of *E. coli* [4] as a basis, we have developed a framework for analyzing the stability of hybrid cell models [5] as well as a technique for sensitivity analysis in this class of models using an extension of classical Metabolic Control Analysis [6]. Currently, we are extending a statistical mechanics method for parameter estimation [7] to our models.

References

[1] Castellanos, M., D.B. Wilson, M.L. Shuler. (2004) A modular minimal cell model: purine and pyrimidine transport and metabolism. PNAS 101:6681-6686

[2] Castellanos, M., K. Kushiro, S.K. Lai, and M.L. Shuler. 2007. A genomically/chemically complete module for synthesis of lipid membrane in a minimal cell. Biotechnol. Bioeng. 97:397-409

[3] Forster, A.C., G.M. Church. (2006) Towards Synthesis of a Minimal Cell. *Molecular Systems Biology* 2, 45.

[4] Shuler, M.L. (2005) Computer Models of Bacterial Cells to Integrate Genomic Detail with Cell Physiology. Proceedings of the KBM International Symposium on Microorganisms and Human Well-Being, Seoul Korea.

[5] Nikolaev, E.V., J.C. Atlas, M.L. Shuler. (2006) Computer Models of Bacterial Cells: From Generalized Coarse-Grained to Genome-Specific Modular Models. *Journal of Physical Conference Series* **46**, 322-326.

[6] Nikolaev, E.V., J.C. Atlas, M.L. Shuler. (2007) Sensitivity and control analysis of periodically forced reaction networks using the Green's function method. J. Theoretical Biol. **247**:442-461

[7] Brown, K.D., J.P. Sethna. (2003) Statistical Mechanical Approaches to Models with Many Poorly Known Parameters. *Phys. Rev. E.* **68**

Acknowledgements: This work was funded by DOE grant DE-FG02-04ER63806 and by NYSTAR, the New York State Office of Science, Technology, and Academic Research. Jordan Atlas gratefully acknowledges support from the DOE Computational Science Graduate Fellowship Program of the Office of Science and National Nuclear Security Administration in the DOE under contract DE-FG02-97ER25308.

439c Robustness in Biological Networks: From Genes to Cells

Francis J. Doyle III, Department of Chemical Engineering, University of California - Santa Barbara, Santa Barbara, CA 93101-9611

A property of particular interest in systems biology is the robustness of a biophysical network: the ability to maintain some target level of behavior or performance in the presence of uncertainty and/or perturbations. In biological systems, these disturbances can be environmental (heat, pH, etc.) or intrinsic to the organism (changes in kinetic parameters arising from molecular fluctuations). While preliminary results are available for simple (low-dimensional, deterministic) biological systems, general tools for analyzing these tradeoffs are the subject of active research.

Tools from systems theory are introduced that elucidate design principles in these complex architectures through the analysis of robust and fragile regions of the network. Analysis of the performance properties of circadian gene networks in Arabidopsis, Drosophila, and Mouse reveals the design principles that emerge from these richly layered and hierarchical regulatory circuits.

We highlight some recent results that analyze robustness properties at the tissue level, where intercellular coupling appears to be responsible for the generation of robust rhythms in the face of noise and other uncertainties. In the absence of intercellular signaling, the individual (cellular) oscillators lose these properties of robust performance.

Finally, we will show that these insights (robustness and performance analysis) can be applied to the design of "forcing" or control protocols that allow for rapid re-entrainment of the clock.

439d Understanding Adaptive Immunity: A Crossroad of the Physical, Life, and Engineering Sciences

Arup K. Chakraborty, Chemical Engineering, Massachusetts Institute of Technology, Cambridge, MA 02139

Higher organisms, like humans, have an adaptive immune system that can combat pathogens that have not been encountered before. The adaptive immune system can also go awry, and many autoimmune disorders (e.g., multiple sclerosis, type I diabetes, etc.) are the direct consequence of the adaptive immune system attacking the organism's own tissues. T lymphocytes (T cells) are the orchestrators of the adaptive immune response. They interact with cells, called antigen presenting cells, which display molecular signatures of pathogens on their surface. T cells detect the presence of these molecular signatures of pathogens with great sensitivity. How T cells hunt for antigen, how they discriminate between "self" and "non-self" with extraordinary sensitivity, and how their activation is regulated and mis-regulated are central questions in fundamental biology. In spite of some spectacular advances, an understanding of the principles that govern how T cell responses are regulated and misregulated has not emerged. A principal reason for this is that the biology of T cells is an example of "emergent complexity"; i.e., the pertinent dynamic events cannot be parsed in terms of a linear superposition of the consequences of binary interactions between the relevant components and the cooperative processes span a spectrum of time and length scales. This hierarchical cooperativity, with feedback, makes it difficult to intuit underlying mechanisms from a few observables. I will discuss recent efforts where bringing together approaches from the physical, life, and engineering sciences has shed light on the mechanisms that regulate this complexity. I will especially highlight how stochastic fluctuations and spatial organization of components can influence the behavior of this system in important ways.

497a Identifying New Biochemical Pathways And Regulatory Elements Using Time-Series Metabolomic Analysis

Harin H. Kanani and Maria I. Klapa. Chemical and Biomolecular Engineering, University of Maryland, College Park, 2113 Chemical and Nuclear Engineering Building, College Park, MD 20742-2111

Detailed understanding of biochemical pathways and common regulatory elements is a pre-requisite for creating system wide *in-silico* models and metabolic engineering of biological systems. Such information for complex eukaryotic and novel prokaryotic systems, is traditionally obtained by (a) post gene-sequencing homology based predictions (b) rigorous gene-by-gene or pathway-by-pathway studies. While the former approach is unable to identify presence of novel pathways and regulatory elements specific to the system being studied, the later approach is not high-throughput. A high-throughput experimental design methodology for identifying novel biochemical networks and regulatory elements would be thus, a very important tool for engineering complex eukaryotic and non- "model prokaryotic" systems: such as mammalian system, new species of filamentous fungi, and not-fully-sequenced but commercially important plants such as switch grass, woody plants, vegetable plants, cotton, etc.

Metabolomics – the newest of the "-omic" technologies, provides a high-throughput, organism independent, tool for obtaining the metabolomic fingerprint of the biological system being studied. In order to use metabolomic analysis for identifying novel biochemical pathways and regulatory elements in whole plant systems we obtained time-series (reference + eight time points in each series) metabolomic profiles of response of 12 day old Arabidopsis thaliana seedlings grown in liquid cultures under normal (control) and seven perturbed conditions. The perturbed conditions were obtained by increasing CO_2 levels, salt stress, sugar (trehalose) signal and hormone (ethylene) signal applied individually and the last three in combination with the first. The polar metabolomic profiles were obtained using Gas Chromatography-Mass Spectrometry (GC-MS) analysis. The obtained raw GC-MS metabolomic profiles were corrected for various derivatization biases using a novel metabolomic data correction algorithm [1] which increased accuracy of metabolomic analysis by an order of magnitude.

Each of these perturbations altered metabolite concentrations significantly, with some metabolites showing up to 30 fold increase / decrease in their relative concentration at individual time points in response to the perturbations. Comparison of response of metabolites belonging to a known metabolomic pathway identified novel regulatory elements of important primary metabolism pathways such as Tri Carboxylic Acid (TCA) cycle and Secondary Amino Acid biosynthesis. In addition by measuring correlations between the time profiles of metabolites under eight different conditions, with successive degree of stringency, we were able to reconstruct parts of known biochemical networks and also identify presence of novel pathways and regulatory elements for A. thaliana. The results thus obtained provide important regulatory insights for plants with important commercial metabolic engineering applications. In addition to providing comprehensive information about the biological system being studied and effects of combination of stress, the present study provides a comprehensive methodology for conducting similar metabolomic studies in other complex eukaryotic or novel prokaryotic systems; getting one step closer to high-throughput system wide reconstruction of their unknown biochemical networks.

[1] Kanani H and Klapa MI. "Data Correction Strategy for Metabolomics Analysis using Gas Chromatography-Mass Spectrometry", Metabolic Engineering, 2007, 9(1):35-51.

Author Present address:

Dr. Harin H. Kanani: Pioneer Hi-Bred International - A DuPont Company, Johnston, Iowa – 50131, USA.

Dr. Maria I. Klapa: Institute of Chemical Engineering and High Temperature Chemical Processes, Foundation of Research and Technology, Hellas (FORTH/ICE-HT), Patras, Greece.

497b Investigation Of Processing And Imputation In Metabolomics Datasets

Lily V. Tong[1], **Mark Styczynski**[1], Joanne K. Kelleher[2], and Gregory Stephanopoulos[1]. (1) Chemical Engineering, Massachusetts Institute of Technology, 77 Massachusetts Ave., 56-439, Cambridge, MA 02139, (2) MIT, 77 Mass Ave, Room 66-401, Cambridge, MA 02139

Metabolomic data contain missing values for a number of reasons, including instrumental limits of detection and improper data alignment between samples. While many supervised and unsupervised data analysis techniques have been used to analyze metabolomic data, the topic of how to process data and handle missing data has largely been ignored. These are well-studied issues in the field of microarrays that have an important parallel in metabolomics datasets.

We surveyed a number of techniques from the microarray literature and determined their applicability to metabolomics data. We used a publicly available GC-MS dataset of *Medicago truncatula* (ten replicates for each sample type of roots, stems, and leaves). We first ran the data through SpectConnect, a metabolomic data analysis tool, requiring 100% metabolite presence in all replicates in order to obtain a complete dataset with no missing values. We then processed the data in three different ways: no data pre-processing, $\log_2$ data transformation, and $\log_2$ transformation with data whitening (zero mean and unit standard deviation).

For each processed data version, we then deleted values at random 1) with a uniform prior distribution and varied total missingness and 2) with a nonuniform prior distribution such that 95% of the missing values are expected to come from some fraction of the smallest GC peaks, also with varied total missingness. Numerous such deletion datasets were created for each processed dataset so as to obtain a satisfactory sample size for performance analysis. We imputed new values for the missing values of each deletion dataset using six prominent techniques from the literature. Error with respect to the complete datasets was calculated based on a normalized root mean square error, and error values for each processing method/imputation method combination were averaged for all iterations at each missingness value.

For the uniform prior distribution, imputation with zeros yielded the highest error for every processing method. The impact of data transformation varied as a function of total data missingness, though general trends were observed that will be described. Depending on data transformation and total missingness, different algorithms may be optimal in imputing missing data. As expected, errors under the nonuniform prior distribution were much higher than those under the uniform prior distribution as the missing data was targeted towards those peaks with the lowest abundance. Trends specific to these circumstances will also be described.

This work has shown that parallels can be drawn between the natures of microarray data and metabolomics data and those techniques used to process and impute missing microarray data can – and perhaps should – be applied to metabolomics data.

497c Protein Turnover Regulates Oxidatively Modified Proteins In Alcoholic Liver Disease: A Proteomic Analysis

Billy W. Newton[1], William Russell[2], Sashi K. Ramaiah[3], and Arul Jayaraman[1]. (1) Chemical Engineering, Texas A&M University, College Station, TX 77843, (2) Chemistry, Texas A&M University, College Station, TX 77843, (3) Veterinary Pathobiology, Texas A&M University, College Station, TX 77843

Chronic disorders and diseases cause profound changes in protein expression levels in affected organ and tissue systems, and proteomic analysis of protein expression can lead to elucidating the mechanisms underlying the disease. Alcoholic Liver Disease (ALD) is a major public health problem and cause for liver dysfunction, however, relatively little is known about the actual mechanisms for the development and progression of ALD. ALD has been associated with increased oxidative stress in the liver and has been shown to cause changes in the expression of proteins associated with antioxidant protection and metabolism. We used two-dimensional gel electrophoresis to characterize murine liver proteomes after 6 weeks exposure to ethanol. Protein changes were quantified using image analysis and protein spots showing significant changes in expression were identified using MALDI-TOF mass spectrometry. A total of 23 spots were identified as differentially expressed between ALD and control liver tissue. The identified proteins included peroxiredoxin 6 (PRXD6) and aldehyde dehydrogenase 2 (ALDH2); however, these proteins were down-regulated in ALD liver by 1.6 and 1.8-fold, respectively. This result is surprising as several studies have shown that chronic alcohol exposure reduces the functional forms of liver enzymes involved in the response to oxidative stress, including PRDX6 and ALDH2. Moreover, since acetaldehyde is a toxic metabolite generated from ethanol, enzymes involved in the further metabolism of acetaldehyde to acetate by ALDH2 were also expected to be up-regulated. Quantitative RT-PCR was used to determine the mRNA levels of these proteins in ALD and control liver tissue to ascertain if changes in the expression of PRXD6 and ALDH2 are seen at the transcriptional level as well. Our data show that PRDX6 and ALDH2 mRNA are either unchanged or even slightly up-regulated in ALD liver tissue, which indicates that changes in the expression of these proteins are post-transcriptionally regulated. Based on our data as well as previous reports suggesting that chronic alcohol exposure results in oxidatively modified proteins (e.g., sulfonation and carbonylation), we hypothesized that the decrease in expression of PRXD6 and ALDH2 in ALD is, in part, due to an increase in the degradation of oxidatively modified proteins.

A proteomic approach has been undertaken to profile the extent of protein carbonylation observed with oxidative stress in an in vitro cell culture model, and test our hypothesis on oxidative stress leading to increased protein turnover. Oxidative stress was induced in cultured HepG2 cells by exposure to hydrogen peroxide (H_2O_2). Protein degradation was inhibited using the molecule mg-132 which inhibits protein turnover through both the ubiquitin/proteasome and lysosomal pathways. Since ALD leads to fat accumulation (steatosis) in hepatocytes, an additional experimental group was included where HepG2 cells were exposed to fatty acids prior to peroxide exposure. Carbonylated proteins from oxidatively stressed HepG2 cells were enriched and separated from unmodified proteins using biotin-streptavidin affinity separation and the level of protein carbonylation quantified. The identities of the carbonylated proteins and the sites of carbonylation are being identified using LC-MS/MS mass spectrometry to develop a fundamental understanding of oxidative changes and protein turnover in ALD.

497d Quantitative Proteomics As A Tool For Systems Biology: Assessment Of Metal Stresses In Burkholderia Cepacia

Carla M. R. Lacerda and Kenneth F. Reardon. Department of Chemical and Biological Engineering, Colorado State University, 1370 Campus Delivery, Fort Collins, CO 80523-1370

Proteomics is a powerful tool for experimental systems biology, relating global protein expression to phenotype. In this project, at strain of *Burkholderia cepacia* was isolated from a soil microbial community due to its high ability to tolerate cadmium. In previous studies, we used proteomics to determine the effect that cadmium shocks had on the whole bacterial community and to evaluate the effect of different cadmium concentrations on the isolate.

In the present study, *B. cepacia* was grown with and without 100 mg/L Cd^{2+}. Cadmium-grown cells presented a very distinct blue phenotype compared to the control. Both types of cells were transferred to new plates containing Co^{2+}, Cr^{3+}, Cu^{2+}, Fe^{3+}, Ni^{2+}, or Zn^{2+}. The objective was to determine whether cells adapted to cadmium would grow faster on the different metals. After approximately 10 days of incubation, no significant difference in growth rates was noted , but new phenotypes were observed.

The overall goal of the experiment was to elucidate mechanisms involved in the protection against high concentrations of metals. Due to the different chemical properties of the metals chosen, different cellular machineries should be involved in metal resistance. Using a proteomic approach to implement a systems view of metal resistance, we had the ability to further describe specific mechanisms used by *B. cepacia*. In order to obtain quantitative information, we coupled iTRAQ labeling with two-dimensional chromatography and quadrupole time-of-flight mass spectrometry. The results reveal remarkable differences in protein expression among the samples, consistent with the previous observation of different phenotypes. Among the proteins identified as differentially expressed were a number of central carbon metabolism enzymes. As a consequence, we tried to gain further insight into the central cellular metabolism by measuring substrate consumption and ATP levels, and calculating cellular yields. The combination of metabolite and protein data offers unique information for future modeling efforts. This study illustrates the ability of proteomics to provide systems information on unknown cellular states.

497e Anaerobic Fermentation Of Glycerol By E. Coli: A Proteomics Study

Ashutosh Gupta and Ramon Gonzalez. Chemical and Biomolecular Engineering, Rice University, MS-362, 6100 S Main St, Rice University, Houston, TX 77005

Glycerol is an abundant and inexpensive carbon source generated as inevitable by-product during the production of biodiesel. The higher reduced state of carbon in glycerol makes it an attractive target for the production of reduced chemicals such as ethanol, succinate, propanediols, etc. To fully realize this potential the use of anaerobic fermentation is required. Fermentative metabolism of glycerol, however, is restricted to very few microorganisms. E. coli, considered the platform of modern biotechnology, was thought for a longtime to metabolize glycerol only via respiration. Unlike previous reports, recent studies in our laboratory indicate that E. coli can anaerobically ferment glycerol (Biotechnol. Bioeng. 94 (5) 821, 2006). To further characterize this metabolic process we compared glycerol fermenting and non-fermenting cells using proteomics technology.

In this study, wild type E. coli MG1655 was grown at slightly acidic conditions, with only half point pH difference (pH 5.5 and 6) between the two cultures. This small change in pH made the cells to go from glycerol non-fermenting (pH 5.5) to glycerol fermenting (pH 6) state. To further identify the differences in metabolic pathways involved in the two conditions, 2D gel electrophoresis coupled with MALDI/TOF technology was used to identify the proteins, which were differentially expressed between the two conditions. Most of the differentially expressed proteins were from the central carbon metabolism of E. coli. Among these proteins, Tryptophanase and pyruvate kinase A were found to be expressed more than five times at the pH 6 condition as compared to pH 5.5. To further understand the metabolic pathways involved in the anaerobic fermentation of glycerol by E.coli, we are also studying some of the E.coli mutants using proteomics tools.

497f A Genome-Scale Analysis Of The Translational Efficiencies Of E. Coli Genes

Hermioni Zouridis, Chemical and Biological Engineering, Northwestern University, 2145 Sheridan Rd., Evanston, IL 60208 and Vassily Hatzimanikatis, Laboratory of Computational Systems Biotechnology, EPFL, CH-1015, Lausanne, Switzerland.

Translation is a central cellular process and the sheer complexity of its mechanism necessitates mathematical frameworks to better understand system properties and make quantitative predictions for many problems in medicine and biotechnology. Central to the translation mechanism is the elongation phase, during which the ribosome facilitates assembly of the polypeptide chain with one amino acid added at each codon along the length of the mRNA. Amino acids are delivered to the ribosome by tRNAs that serve as adapter molecules between the amino acid and the codon occupying the ribosomal A site.

Because of the important role codons play in translation kinetics, codon usage patterns have been correlated with gene characteristics such as expression level, intragenic position, and length. However, codons have varying elongation kinetics due to different tRNA availabilities, codon – anticodon compatibilities, and the multiple elementary steps and translational components involved in the elongation cycle at every codon. Hence, in order to better understand the relationship between codon usage patterns and protein synthesis properties, we have developed a gene sequence specific mechanistic model for translation which accounts for all the elementary steps of translation elongation.

We apply our mechanistic framework to the translation of every protein in E. coli in order to determine the translational efficiency of each gene. We also perform a sensitivity analysis to determine the contribution of elongation kinetic parameters of each codon on the overall protein synthesis rate of each gene. We observe that the relative position of codons along the mRNA determines the rate limiting effect of the individual codons and the optimal protein synthesis rate. Applying our mechanistic framework to E. coli genes aids in understanding the complex, nonlinear interplay between codon usage and protein synthesis properties, and in characterizing how these properties relate to patterns such as gene expression levels and function in cells. Based on these studies we propose some principles for the design of vectors for optimal protein production.

538a A New Optimization Based Approach For Protein Residue Contact Prediction

Rohit Rajgaria[1], Scott R. McAllister[1], and Christodoulos A. Floudas[2]. (1) Chemical Engineering, Princeton University, Dept of Chemical Engineering; A215 Engineering Quadrangle, Princeton, NJ 08544, (2) Department of Chemical Engineering, Princeton University, Engineering Quadrangle, Princeton, NJ 08544

Predicting the residue contacts of a protein is an important component of protein structure prediction. It particularly plays an important role in first principles based approaches where no database information is used for the secondary and tertiary structure prediction. Residue contact prediction mainly focuses on the prediction of residue pairs which are far apart in the primary sequence of a protein but proximal in their three dimensional structure. A number of different methods have been developed to predict these non-local contacts. In a very broad sense, these techniques either use correlated mutations analysis or machine learning approach (hidden Markov models, neural networks etc.) [1-5].

We have recently developed an optimization based method to predict the residue contacts of a protein. In this model, a Cα- Cα distance dependent force field [6] is used to assign contact energy for a particular contact based on the identity of the amino acids [7]. A contact is said to occur when the predicted distance between interacting residues is more than 3 •, and less than 9 •. This problem has been formulated as a mixed integer linear programming problem where the objective function is to minimize the contact energy. A set of constraints is also included in the model to produce physically possible contacts. Apart from predicting the non-local contacts, this method can also predict the topology of a protein and disulfide bridges (if present). This formulation not only offers the advantage of finding the residue contacts corresponding to the global minima, but it can also produce a rank-ordered list of residue contacts. A rank-ordered list like this helps in finding the most frequent contacts. It is also possible to encounter cases when some of the contacts are known a priori. The proposed formulation allows the user to add such problem specific biologically relevant information in a mathematical rigorous way. This model has been tested on a test set including α, β, and α+ β protein and the preliminary results are very encouraging.

[1] Klepeis, J., and Floudas, C. A., 2003a, Prediction of Beta-Sheet Topology and Disulfide Bridges in Polypeptides. Journal of Computational Chemistry, 24: 191-208.

[2] Vicatos, S., Reddy, V. B., and Kaznessis, Y., 2005, Prediction of Distant Residue Contacts With the Use of Evolutionary Information. Proteins: Structure, Function, and Bioinformatics, 58: 935-949.

[3] Fariselli, P., Olema, O., Valencia, A., and Casadio. R, 2001, Progress in Predicting Inter-Residue Contacts of Proteins With Neural Networks and Correlated Mutations. Proteins: Structure, Function, and Genetics, 5: 157-162.

[4] Hamilton, N., Burrage, K., Ragan, M. A., and Huber, T., 2004, Protein Contact Prediction Using Patterns of Correlation. Proteins: Structure, Function, and Genetics, 56: 679-684.

[5] Cheng, J., and Baldi P., Improved Residue Contact Prediction Using Support Vector Machines and A Large Feature Set. BMC Bioinformatics, 56: 679-684.

[6] Rajgaria, R., McAllister, S. R., and Floudas, C.A., 2007, Development of A Novel High Resolution Cα- Cα Distance Dependent Force Field Using A High Quality Decoy Set. Proteins: Structure, Function, and Bioinformatics, 65: 726-742.

[7] Rajgaria, R., McAllister, S. R., and Floudas, C.A., 2007, In preparation.

538b A New Group Contribution Method For Genome-Scale Thermodynamic Analysis

Christopher S. Henry[1], Linda J. Broadbelt[1], and Vassily Hatzimanikatis[2]. (1) Northwestern University, 2145 Sheridan Road, E136, Evanston, IL 60208-3120, (2) Laboratory of Computational Systems Biotechnology, EPFL, CH-1015, Lausanne, Switzerland

As metabolic models grow larger and more complex and methods for the thermodynamic analysis of these models emerge, knowledge of the Gibbs free energy of formation ($\Delta_f G'^{\circ}$) and reaction ($\Delta_r G'^{\circ}$) for the compounds and reactions in these models is becoming essential. We introduce an expanded version of the group contribution method for the estimation of $\Delta_f G'^{\circ}$ and $\Delta_r G'^{\circ}$ in biochemical systems. Gibbs energy contributions are fit using 74 distinct molecular substructures and 11 interaction factors against a training set of 644 reactions and 224 compounds. We utilized bootstrap analysis to determine the accuracy of the methodology in estimating $\Delta_r G'^{\circ}$ and $\Delta_f G'^{\circ}$ for reactions and compounds not included in the training set of the fitting. Based on the results of the bootstrapping, the standard error involved in these estimations was 2.09 kcal/mol.

The new group contribution method is applied to estimating $\Delta_f G'^{\circ}$ and $\Delta_r G'^{\circ}$ for the metabolites and reactions in a genome-scale metabolic model of *E. coli*. The new estimated $\Delta_f G'^{\circ}$ and $\Delta_r G'^{\circ}$ values were utilized in a Thermodynamics-based Metabolic Flux Analysis (TMFA) of *E. coli* metabolism (1). TMFA is a form of metabolic flux analysis that includes thermodynamic constraints in addition to the mass balance constraints of metabolic flux analysis. Using TMFA we determined the thermodynamically feasible ranges for the reaction fluxes, reaction , and metabolite activities in the *E. coli* metabolism. The results indicate that most reactions essential for growth in *E. coli* have a sufficiently favorable and the physiological range of metabolite activities is sufficiently flexible to allow thermodynamic feasibility of these reactions under a wide range of conditions. A small number of reactions emerge as significant thermodynamic bottlenecks in the metabolic model, and of these reactions and the activities of the metabolites involved in these reactions are constrained to narrow ranges. Overall, the *E. coli* metabolism was found to be remarkably resilient to the thermodynamic constraints.

1. Henry, C. S., L. J. Broadbelt and V. Hatzimanikatis. 2007. Thermodynamics-based metabolic flux analysis. Biophys J. 92(7):1792-1805.

Web Page: systemsbiology.northwestern.edu/

538c Determination Of Viral Infection Rate Constants Through Computational Analysis

Jacob McGill and Mariajose Castellanos. Chemical and Biochemical Engineering, UMBC, 1000 Hilltop Circle, Baltimore, MD 21250

Well developed structured models can tell us a great deal about the dynamics of viral infection within cells. Models of this nature can lead to better antiviral drugs by revealing where the weakest links are in viral synthesis. Unfortunately, many difficulties can arise when creating a structured model, including obtaining rate constants and concentration data for each step and finally validating the mechanism of infection. We have developed a structured stochastic simulation to test some of these issues. Influenza A was chosen since its infection mechanism is relatively well documented and there is large amount of published data on its various rate constants; although there are many values that remain ambiguous. Using various statistical experiments and response surface models of all the constants in the simulation, over 45 in total, a "best guess" at the remaining constants was made. Under the assumption that nature tends toward optimizing its processes to edge out competition, the optimization of the response surface models given various constraints should result in calculated rate constants that closely mimic the 'true' constants. This methodology can be applied to the mechanism itself. It stands to reason that if the proposed mechanism used in the simulation cannot be solved with given real world constraints then an alternate mechanism should be proposed. Additionally, experimentally determined rate constants can also be used to validate the optimized results.

538d Optimal Experimental Design For Parameter Estimation In Biological Signal Transduction Modeling

Marc R. Birtwistle[1], Boris N. Kholodenko[2], Jan B. Hoek[2], and Babatunde A. Ogunnaike[1]. (1) Chemical Engineering, University of Delaware, 150 Academy St., Newark, DE 19716, (2) Pathology, Anatomy, and Cell Biology, Thomas Jefferson University, 1020 Locust St., Philadelphia, PA 19107

By providing quantitative insight into signal transduction pathophysiology, systems biology promises to revolutionize the pharmaceutical industry. However, the insight provided to date by current modeling approaches have yet to show substantial impact on real-world industrial and medical applications. We believe that an important contributing factor to this lack of translational impact is the intrinsic difficulty in estimating the parameters for such models uniquely and accurately, since such models typically contain on the order of 100 non-linear ordinary differential equations (ODE) and twice as many unknown parameters. A main difficulty in parameter estimation, in addition to the sheer size of the parameter search space, is that there is no systematic procedure for evaluating parameter identifiability, that is, which parameters can be uniquely estimated from a particular set of experimental data. As a result, there are currently no methods for designing experiments (what to measure and at what time point, using what perturbations) for the purposes of parameter estimation in such models. Addressing these fundamental modeling problems, we present in the current work metrics for evaluating parameter identifiability in signal transduction models, and use these metrics to develop a methodology for optimal experimental design (OED) for parameter estimation in biological signal transduction modeling. We apply this methodology to two well-known models of mammalian signal transduction involving short-term Epidermal Growth Factor (EGF) signaling: the Kholodenko (1999) and Schoeberl (2002) models. First, we show that with the experimental data reported in these studies, the parameters that were estimated are *not* uniquely identifiable. Second, we propose how these unknown parameter values can be identified using OEDs based on traditional, low-throughput measurements of downstream signaling in response to EGF stimulation, with and without various pharmacological inhibitors. Third, we explore the OEDs when high-throughput, quantitative proteomic measurement of signaling responses are assumed to be readily available, showing that these measurements have the potential to be much more efficient than traditional methods for parameter estimation in this class of models. Overall, our method lays a foundation for effective experimental design and parameter estimation in biological signal transduction modeling, and is generally applicable to high-dimension, non-linear ODE modeling.

References

Kholodenko, B.N., Demin, O.V., Moehren, G. and Hoek, J.B. (1999) Quantification of short term signaling by the epidermal growth factor receptor. *J Biol Chem*, **274**, 30169-30181.

Schoeberl, B., Eichler-Jonsson, C., Gilles, E.D. and Muller, G. (2002) Computational modeling of the dynamics of the MAP kinase cascade activated by surface and internalized EGF receptors. *Nat Biotechnol*, **20**, 370-375.

538e Seasonal Adaptation By Coupled Morning And Evening Circadian Oscillators

Robert S. Kuczenski, School of Chemical and Biomolecular Engineering, Cornell University, Ithaca, NY 14853-5201 and Kelvin H. Lee, Department of Chemical Engineering, University of Delaware, 150 Academy Street, Colburn Laboratory, Newark, DE 19716.

Circadian rhythmicity allows organisms to adapt to regular day-night cycling and, importantly, to seasonal changes in their relative lengths. Recent experimental observations in the fly *Drosophila* have identified an additional class of neurons, the Evening-oscillators (E-cells) which when coupled with the well-characterized Morning-oscillators (M-cells) allow seasonal adapation to day length. Bulding upon our group's previously created mathematical model of the *Drosophila* circadian network in M-cells, we first study the gene network modifications necessary to create the experimentally observed behavior of the closely related E-cells. With a characterized E-cell gene network, we then stochastically simulate a small number of M- and E-cells to study the possible coupling mechanisms that yield the experimentally observed day length-dependent domination of one cell type over the other. This study will complement experimental efforts by providing both further insight into the possible mechanisms underlying the entrainment of circadian rhythms and the adaptation to widely varied seasonal changes in day length.

538f Deterministic And Stochastic Modeling Of A Caricature Model For The Cell Cycle

Michail Stamatakis and Nikos Mantzaris. Chemical and Biomolecular Engineering Department, Rice University, Houston, TX 77005

Deterministic formulations have been extensively used over the past decades to model the cell cycle. This approach, however, is inadequate for intracellular reaction phenomena occurring in control volumes on the order of some pico-liters: the probabilistic fashion in which reactions occur in every single cell creates significant random fluctuations, termed "intrinsic noise". Therefore, the underlying dynamics of transcription and translation, the protein-protein and protein-ligand interactions, as well as the degradation phenomena, must be described in probabilistic terms, with the use of stochastic models. Taking into account this inherent intracellular stochasticity can reveal patterns in the behavior of a biochemical network that are different from those predicted by a deterministic model. In this work, we have developed a reaction network that mimics the salient processes involved in cell cycle. The network incorporates a positive and a negative feedback loop: the positive feedback stems from the activation of a cyclin-dependent kinase by a cyclin and the subsequent autocatalytic effect in the expression of cyclins. The negative feedback arises from the promotion of S-phase inhibitors' degradation by cyclin-CDK complexes. From the reaction network we derive equivalent deterministic and stochastic models and compare the predictions of the two in order to reveal the effect of stochasticity in the behavior of the system. Comparisons are performed in different regions of the parameter space, using a bifurcation parameter that quantifies the strength of the positive feedback. The comparisons reveal that stochasticity can convert deterministic excitability to stochastic oscillations thereby increasing the range of the bifurcation parameter where the system exhibits oscillatory behavior. This effect may have a physiological importance since oscillations are essential for the function of the cell cycle.

Web Page: www.ruf.rice.edu/~nman

538g A Mathematical Model Of Acute Inflammatory Response To Endotoxin Challenge

Anirban Roy[1], Gilles Clermont[2], Silvia Daun[2], and Robert S. Parker[3]. (1) Department of Chemical and Petroleum Engineering, University of Pittsburgh, 3700 O'Hara Street, Room 1249 Benedum Hall, Pittsburgh, PA 15261, (2) Critical Care Medicine, University of Pittsburgh, 606B Scaife Hall, Pittsburgh, PA 15261, (3) Chemical and Petroleum Engineering, University of Pittsburgh, 3700 O'Hara Street, Room 1249 Benedum Hall, Pittsburgh, PA 15261

INTRODUCTION: The acute inflammatory response system is responsible for the initial action of the body against acute biological stress, such as bacterial infection (endotoxin) or tissue trauma. Such response action involves a cascade of events mediated by an array of cells and molecules to eliminate invading pathogens and repair damaged tissues [1]. For a normally functioning response system, the inflammation eventually subsides and the body returns to basal level. However, an excessive inflammatory response can lead to further tissue damage, organ dysfunction, or possibly death [2]. To control the inflammatory response system, organisms have developed regulatory mechanisms such as pro- and anti-inflammatory mediators. In general, the pro-inflammatory elements (e.g., interleukin-6 (IL-6), tumor necrosis factor (TNF)) up-regulate the inflammatory actions; whereas, the anti-inflammatory elements (e.g., interleukin-10 (IL-10)) down-regulate the inflammation response [3]. In this study, an ordinary differential equation (ODE) model of the acute inflammation response was developed to capture the dynamics of inflammatory cells and the dynamics and interactions of various pro- and anti-inflammatory mediators.

METHODS: The mathematical model consists of 8 ODEs and 42 parameters. The dependent variables include the concentration of endotoxin challenge (P), concentration of activated phagocytic cells (N), tissue damage marker (D), pro-inflammatory mediator concentrations (IL-6 and TNF), IL-10 concentration (which is an anti-inflammatory cytokine), and other anti-inflammatory moderator concentrations (Ca). The differential equations include biologically-motivated saturation terms (e.g., Hill functions), thus making the model nonlinear. Values for some of the parameters were obtained from A. Reynolds et al [4]. In order to calibrate the model, data was obtained from experiments performed on Sprague-Dawley rats by Dr. Y. Vodovotz and colleagues at the University of Pittsburgh (Department of Surgery). Model parameters were estimated using experimental data of endotoxin challenges at 3 and 12 mg/kg via the nonlinear least square technique [5]. Finally model validation was performed by comparing the model predictions at an endotoxin challenge level of 6 mg/kg with experimental data at the same level.

RESULTS: The introduction of an endotoxin challenge (P) activates the phagocytic cells (N), which in turn, up-regulates all the mediators (TNF, IL-10, IL-6, and Ca). Furthermore, the mediators have a positive or a negative feedback based on their functionality (pro- or anti-inflammatory) on the process. The sepsis model successfully captured the dynamics of pro-inflammatory (IL-6 and TNF) and anti-inflammatory (IL-10 and Ca) mediators during endotoxin challenge to rat at dose levels of 3, 6, and 12 mg/kg. Model predictions of the activated phagocytic cells (N) in response to endotoxin challenges at the same dose levels were also consistent with the experimental data.

CONCLUSION: The model successfully emulated the acute inflammatory response to endotoxin challenges at different dose levels. This 8-state model may provide further insight of the various interactions between phagocytic cells and pro/anti-inflammatory mediators. With further validation, the model may provide more information regarding the critical levels of various elements of the inflammatory response system which could lead to sepsis. Finally, this model is a test-bed on which control strategies can be tested. One formulation is model-based control of inflammatory mediators in patients under biological stress by adjusting blood flow through an external recirculation adsorption column.

REFERENCES:

[1] J. Day, J. Rubin, Y. Vodovotz, C.C. Chow, A. Reynolds, and G. Clermont. A reduced mathematical model of the acute inflammatory response. II. Capturing scenarios of repeated endotoxin administration. J. Theor. Biol., 242: 237 - 256, 2006

[2] R.J. Goris, T.P. te Boekhorst, J.K. Nuytinck, J.S. Gimbrere. Multiple-organ failure. Generalized autodestructive inflammation? Arch. Surg., 120: 1109 - 1115, 1985

[3] C. Nathan. Points of control in inflammation. Nature, 420: 846 - 852, 2002

[4] A. Reynolds, J. Rubin, G. Clermont, J. Day, Y. Vodovotz, and G.B. Ermentrout . A reduced mathematical model of the acute inflammatory response. I. Derivation of model and analysis of anti-inflammation. J. Theor. Biol., 242: 220 - 236, 2006

[5] E. Carson, C. Cobelli. Modelling methodology for physiology and medicine. San Diego, CA; Academic Press, 2001

594a Development of Validated Methods for Initial Value Problems Using a Constraint Satisfaction Approach

Ozlem Yilmaz[1], Luke E. K. Achenie[2], and Ranjan Srivastava[1]. (1) University of Connecticut, 191 Auditorium Rd. Unit 3222, Storrs, CT 06269, (2) Department of Chemical Engineering, Virginia Polytechnic Institute and State University, Randolph Hall 133, Blacksburg, VA 24061

ODE systems describing chemical, biological and physical systems usually contain parameters that are only approximately known. Rather than describing these parameters as having point values, they are more accurately described as falling within a range of values. However, to determine the solution space for a range of parameter values requires an infinite number of computer runs, since every point in the range has to be sampled. Interval methods, also referred to as validated methods, for initial value ODE problems (IVPs) are guaranteed to enclose solutions arising from any combination of ranges of parameter values. One approach to addressing the parameter estimation problem is to use the constraint-satisfaction approach where the difference between the model data and the experimental data is employed as the constraint. The aim of this work was to achieve parameter estimation using such a method. As a case study, we looked at the mathematical model that explains how lytic RNA phage infects Escherichia coli C-3000. The system under study described viral dynamics between the phage and its host at the intercellular level via the following variables: uninfected cell density (sensitive and resistant type), infected cell density, free phage density and substrate (glucose) concentration [1]. Model equations denoted by IVPs were converted into a set of polynomial algebraic equations using Taylor model integration methods. The final set of equations were solved by an interval solver, RealPaver [2], employing constraint satisfaction techniques.

References: 1) R. Jain, A.L. Knorr and R. Srivastava, Investigation of Bacteriophage MS2 Viral Dynamics Using Model Discrimination Analysis and the Implications for Phage Therapy, Biot. Prog. 22 (6): 1650-1658. 2) L. Granvilliers and F. Benhamou, RealPaver: An Interval Solver Using Constraint Satisfaction Techniques, ACM Trans. On Math. Software, 32 (2006) 138-156.

594b Multi-Scale Models For Synthetic Bioengineering

Yiannis N. Kaznessis, Chemical Engineering and Materials Science, University of Minnesota, 151 Amundson Hall, 421 Washington Ave SE, Minneapolis, MN 55454

Biological engineering is emerging from biology, as a distinct discipline based on quantification. This quantification of biological phenomena is necessary to rationalize engineering efforts without solely resorting to the traditional description-based paradigm of biological sciences.

Mathematical modeling helped chemical engineering in the 1950s to develop its unique, successful identity. In a similar way, mathematical theories that reduce biological knowledge are becoming sine qua non for progress in biological engineering.

Although the principles of thermodynamics and kinetics apply to biological systems, these systems differ from industrial-scale chemical systems in a fundamental way: they are far from the thermodynamic limit. This hinders the application of mathematics developed by chemical engineers to model kinetic and thermodynamic processes in living organisms.

For example, there are networks of biomolecular interactions (e.g. gene regulatory networks) whose kinetic behavior is key for emerging phenotypes. Models of these interactions are important for rationalizing genetic engineering of DNA sequences that lead to targeted phenotypes. Using deterministic, ordinary differential equations for simulating the reaction kinetics assumes the absence of thermal, stochastic noise in biomolecular systems. This can indeed be distinctly false.

We will explore the limitations of traditional mathematical frameworks and we will discuss emerging models that account for the distinct nature of biological organisms. We will detail the multi-scale algorithms we developed to properly capture biological complexity from the molecular to the system level.

We will describe how these systems models are being used in synthetic bioengineering of novel gene regulatory networks, such as a bio-logical AND-gate synthetic promoter, designed and characterized in silico, cloned in E.coli and tested in our lab.

Web Page: www.cems.umn.edu/research/kaznessis

594c An Optimization Based Method To Automatically Reconcile In Silico/in Vivo Growth Predictions

Vinay S. Kumar, Department of Industrial Engineering, The Pennsylvania State University, 147A, Fenske Lab, University Park, PA 16802 and Costas D. Maranas, Department of Chemical Engineering, The Pennsylvania State University, University Park, PA 16802.

The rapid pace at which genome scale reconstructions are becoming available has necessitated the need to validate in silico growth predictions with in vivo data. One way of carrying out this validation is by contrasting in silico predicted growth or lackthereof with in vivo experimental data for different gene knockouts and/or different carbon sources. Specifically, by adopting an FBA based description the biomass formation is maximized subject to stoichiometry, uptake scenario and possible gene knock-outs. If the maximum flux into biomass is found to be zero then it is deemed that the in silico prediction is for no-growth (i.e., NG). Alternative if the max flux into biomass is non-zero then the corresponding in silico prediction is for growth (i.e., G). These in silico prediction are then contrasted against the in vivo experimental prediction (i.e., G or NG). This yields four possible cases: (i) GG when both model and experiment predict growth; (i) GNG when the model predicts growth but the in vivo prediction is for no-growth, (iii) NGG when the model predicts for no growth but there in vivo growth, and finally (iv) NGNG when both model and experiment predict no growth. The idea is to try to automatically reconcile GNG predictions (by removing reactions from the model) and NGG predictions (by adding reactions to the model). A bilevel optimization formulation and solution procedure is put forth for minimally resolving all observed inconsistencies. The approach is demonstrated using the E. coli model.

594d From Molecular Networks To Multicellular Structures: Predicting Phenotype Diversity And Transitions

Anand R. Asthagiri and Claudiu Giurumescu. Chemical Engineering, California Institute of Technology, 1200 E. California Blvd MC 210-41, Pasadena, CA 91125

Multicellular patterns and structures emerge as cells execute instructions received from cues in their microenvironment. Deciphering how cells integrate these cues to achieve an organized, functional structure is a fundamental question in biology with important biomedical implications in areas such as tissue engineering and regenerative medicine. An essential element in multicellular assembly involves direct cell-cell interactions. Cell-cell contacts are not only physical links between neighboring cells, but also sources of biochemical signals that instruct cellular behavior. Our lab has been focused on elucidating how direct cell-cell interactions crosstalk with other environmental cues to affect individual cell fate choices, and thereby, generate multicellular patterns.

This talk will focus on our recent work in modeling multicellular patterning during the development of the organism, C. elegans. An intriguing aspect of this system is that patterning is guided by both a soluble signal and direct cell-cell interactions. Our computational analysis reveals that cell-cell coupling enhances cell perception of a gradient in the soluble signal. In addition, we have developed a parameter-unbiased, computational framework that accurately predicts wild-type and mutant phenotypes. This ability to predict phenotypes has opened the door to intriguing new questions such as: what new phenotypes are possible, and what perturbations (mutations) render such phenotypes? I will describe how answering these questions have offered a unique opportunity to explore both the molecular genetics of patterning and the evolutionary capacity of this signaling network for creating phenotypic diversity.

Web Page: www.che.caltech.edu/groups/ara/

594e Scaffold-Mediated Temporal Control of Cell Signaling Dynamics

Arup K. Chakraborty, Chemical Engineering, Massachusetts Institute of Technology, Cambridge, MA 02139 and **Jason W. Locasale**, Biological Engineering, Massachusetts Institute of Techonology, 50 Ames St, E19-534, Cambridge, MA 02139.

The ubiquity of scaffold proteins in cell signaling pathways suggests that scaffold proteins are likely to have many regulatory functions. Here, we present Monte Carlo simulations of a model kinase cascade to investigate how characteristics of signaling cascades are influenced by the presence of scaffold proteins. Our results indicate that scaffold proteins can have a dual-regulatory function by amplifying and attenuating the amplitude of incoming signals depending on the physiological context. We also study how scaffold proteins can alter the temporal characteristics of a signal propagating along a kinase cascade. These simulations suggest that scaffolds can influence signal propagation by changing the duration of signaling as well how the output signal is distributed over time. Such behavior illustrates one of the many control properties that scaffold proteins may confer to a cellular signaling process. It is our hope that these simulations provide a roadmap for future experiments – several key predictions obtained from our model can be tested with single-molecule technologies.

594f Towards An Integrated Computational Framework To Model Metabolism And The Synthesis Of Protein Products In Mammalian Cell Culture

Adam C. Baughman, Lealon L. Martin, and Susan T. Sharfstein. Chemical & Biological Engineering, Rensselaer Polytechnic Institute, 110 8th Street, Ricketts Building, Troy, NY 12180

We describe the formulation of a novel, multi-level, computational framework to comprehensively model mammalian metabolism from the point of nutrient uptake, through the entire known network of catabolic (energetic) and anabolic (biosynthetic) cellular reactions, and ultimately to the point at which a large protein product is synthesized and possibly secreted. This framework extends core concepts of an established systems modeling approach ("state-space") onto biological systems – a previously unexplored application of this technique. Further, we will demonstrate the ability of this framework to identify valid, yet non-intuitive, metabolic chemistry when tasked with the optimization of cellular processes relative to the maximum yield of a protein product. Finally, to integrate this computational effort with the experimental laboratory, we explore specific challenges in the engineering of cell lines to adopt this non-intuitive metabolic chemistry. It is our hope that such engineered cell lines will demonstrate the improved productivity predicted by our model.

594g Probing The Effects Of Stochasticity In Biochemical Reaction Networks Using Multiscale Simulation

Leonard A. Harris, Aaron M. Piccirilli, Emily R. Majusiak, and Paulette Clancy. School of Chemical and Biomolecular Engineering, Cornell University, Ithaca, NY 14850

The role and consequences of stochastic noise in biological systems continues to be a subject of great interest within theoretical biology and biophysics. In cellular systems, probabilistic interactions between components present in small numbers, such as genes, mRNAs and proteins, result in noisy intracellular dynamics and have been implicated as the source of non-genetic variability in clonal cell populations. From a computational standpoint, such effects can lead to dynamical behaviors that deviate from predictions based on deterministic ordinary differential equations (ODEs). This fact has motivated the use of stochastic simulation methods for studying the dynamics of biochemical reaction networks; Gillespie's "stochastic simulation algorithm" (SSA), in particular, has found widespread use in recent years. Stochastic methods such as the SSA are known to be limited to relatively small system sizes and short time courses, however, thus limiting their utility for studying realistic biological processes. As a result, great effort has been expended recently in developing accelerated stochastic simulation methods, of which so-called "leaping" approaches are particularly popular. Despite their popularity, however, very few biologically relevant applications of leaping have been reported in the literature. This is a curious fact that has yet to be fully explored or explained.

Here, we use our recent contribution, the "partitioned leaping algorithm" (PLA), to investigate stochastic-to-deterministic transitions in two simple, yet biologically relevant, reaction networks. We probe stochastic effects arising in these systems, uncovering subtleties associated with their dynamics and garnering insight as to why leaping methods have not yet found widespread use in computational systems biology. Specifically, reactions with small populations and large rate constants, which are common in biological systems, cause the method to "bog down" by limiting the number of reaction firings possible at each simulation step and minimizing any gains due to leaping. We show how model reduction can be used to help attenuate this problem and demonstrate its use in conjunction with leaping. Overall, our results lend insight into questions regarding the appropriateness of deterministic descriptions for modeling biochemical systems dynamics, the feasibility of using exact-stochastic methods and the role played by model reduction.

636a Molecular Systems Biology Via Multiscale Modeling And High-Performance Computing

Ravi Radhakrishnan, Andrew Shih, Jeremy Purvis, Yingting Liu, and Neeraj Agrawal. Department of Bioengineering, University of Pennsylvania, 120 Hayden Hall, 3320 Smithwalk, Philadelphia, PA 19104

Recent biochemical and epidemiological studies have shown that the signaling through the epidermal growth factor receptor (EGFR) can be sensitive to various tyrosine kinase inhibitors (TKIs) depending on the receptor's expression level and whether or not the tyrosine kinase domain harbors any somatic mutations. The rationale behind the varied response is still unclear, i.e., why certain cell-lines are hyper-sensitive to TKI treatment or how these inhibitors affect cell-wide signaling patterns, in particular, the attenuation of the oncogenic growth signals. We recently developed a hierarchical multiscale computational approach based on molecular dynamics simulations, free energy based molecular docking simulations, deterministic network-based kinetic modeling, and hybrid discrete/continuum stochastic dynamics protocols to study the dimer-mediated receptor activation characteristics, signal transduction, and inhibition of the Erb family receptors, specifically the epidermal growth factor receptor (EGFR) [1]. Extending this previous work, here we predict the inhibitory effects on receptor phosphorylation and downstream signaling response [2]. Our results suggest that the increased drug sensitivity observed for a commonly mutated form of EGFR, L834R, can be attributed to its altered kinetic behavior during receptor phosphorylation. The also model predicts the kinase inhibitor to be more effective in the L834R system under over-expression of EGFR. Additionally, we find that the TKI erlotinib is more effective in inhibiting the Akt response. Our modeling results are compared and validated against "multiscale" experiments ranging from crystallography, phosphorylation kinetics using time-of-flight mass spectroscopy, and cell-based immuno assays with overall qualitative agreement and quantitative agreement where possible. We believe that the multiscale computational framework described here is ideal for assessing mutation landscape on signal transduction. We also believe that our model driven approach will in the long-term significantly impact the optimization of future small molecule therapeutic inhibition strategies as well as the formulation of drug-resistance models.

Y. Liuμ, J. Purvisμ, A. Shih, J. Weinstein, N. Agrawal, R. Radhakrishnan, "A multiscale computational approach to dissect early events in the Erb family receptor mediated activation, preferential signaling, and relevance to oncogenic transformations", 2007, Annals of Biomedical Engineering, in press, DOI: 10.1007/s10439-006-9251-0; μ these authors contributed equally.

J. Purvis, Y. Liu, V. Ilango, and R. Radhakrishnan, Untangling the efficacy of tyrosine kinase inhibitors in the mutants of the epidermal growth factor receptor through a multiscale molecular/ systems model for phosphorylation and inhibition, submitted to IEE Sys Biol.

636b Coarse-Grained Dynamics Of A Neural Field Model

Thomas A. Frewen[1], Carlo R. Laing[2], and Ioannis G. Kevrekidis[1]. (1) Department of Chemical Engineering, Princeton University, Engineering Quadrangle, Princeton, NJ 08544, (2) Institute of Information and Mathematical Sciences, Massey University, Private Bag 102-904 NSMC, Auckland, New Zealand

We study a stochastic spatio-temporal pattern forming system used to model neural activity in the cortex. Neural dynamics are intrinsically noisy, and we investigate the effects of noise on models for pattern formation. In the absence of noise the system dynamics display a single stable activity "bump" that moves around the (periodic) domain in one direction; in the presence of noise the bump occasionally switches direction.

We derive an effective low-dimensional description of the system by data processing (using diffusion maps) the results of short bursts of direct simulation. We utilize this reduced representation, in the form of an effective Langevin equation characterized by a double-well potential, and illustrate that it is quantitatively accurate. This approach bypasses the difficult task of deriving such a potential from the original high-dimensional, stochastic system. Computational bifurcation analysis is also presented based on the coarse-grained description of the system. The noisy discretized system is shown to undergo an effective pitchfork bifurcation. The effective potential and the effects of changing parameters on this potential are also studied.

Finding a low-dimensional description for this system not only enhances our insight into the fundamental dynamics, but also enables us to simulate and analyze this system in a computationally expedient manner. We expect such an approach to be widely applicable to other complex systems.

636c A Multiscale Systems Model For Platelet Signaling And Activation

Jeremy Purvis, Institute for Medicine and Engineering, University of Pennsylvania, 3340 Smith Walk, 1150 Vagelos Laboratories, Philadelphia, PA 19104 and Scott L. Diamond, Department of Chemical and Biomolecular Engineering, University of Pennsylvania, 3340 Smith Walk, 1150 Vagelos Labs, Philadelphia, PA 19104.

Upon activation, platelets release Ca^{2+} from intracellular stores and facilitate Ca^{2+} influx from the extracellular environment, triggering a host of chemical and morphological changes. To gain a better understanding of this process, we have constructed a multiscale, compartmentalized systems model of Ca^{2+} balance in the resting (steady state) and activated (time-dependent) platelet. The model incorporates multiple time and length scales, deterministic and stochastic kinetic simulators, and spatially-resolved depictions of the platelet environment to address specific quantitative questions about platelet function. The resting model comprises five isotropic compartments, 20 chemical species, 27 parameters, and four classes of ODE-based reaction equations: release, sequestration, influx, and extrusion (21 total reactions). Balance between compartments was achieved by the action of a Ca^{2+} ATPase, which counteracts the steady leakage of ions through open IP3 receptor channels by pumping them back against the concentration gradient. Free parameters corresponding to the number of channels and channel activation dynamics were calibrated with experimental data. The system was found to be stable under mild fluctuation, including a 100 nM spike in cytosolic Ca^{2+}, but could be activated by a sufficient increase in key activators, such as extracellular ATP or intracellular IP_3. Using standard electrochemical calculations and measurements for single-channel Ca^{2+} flux, we were able to derive an estimate for the number of Ca^{2+} channels per platelet, which has eluded experimental measurement. Our calculations suggest that as few as 10 open channels per platelet are sufficient to cause the observed burst in intracellular Ca^{2+}. Likewise, we observed that less than 20 IP_3 molecules are sufficient to trigger the same response. These calculations support the view that the heterogeneous signaling behavior observed among platelet populations is partly a stochastic feature due to low molecule copy number. To model signal transduction at the platelet plasma membrane, we developed a spatially-resolved kinetic Monte Carlo procedure which tracks individual interactions among the molecules localized to the lipid bilayer. As a pilot system, we simulated prothrombinase assembly on an 1850 nm^2 patch of the plasma membrane. Explicit modeling of the 2D platelet surface allows allows us to extrapolate physical parameters (e.g., "sticking" coefficients, surface mobilities) that are difficult or impossible to measure experimentally. The pilot surface model comprises seven species and six reactions, including adsorption, desorption, surface association/dissociation, and diffusion. Using fura-2-loaded human platelets and high-throughput robotics, we can explore the combinatorial effect of multiple platelet activators, such as thrombin, ADP and convulxin, as they converge on Ca^{2+} signaling pathways. These data will be used to calibrate a time-dependent model of platelet activation that is consistent with the steady state model. Machine learning will be used to assess the relative contributions of these activation pathways and propose new functional relationships among the network of signaling molecules.

Web Page: bloodsystemsbiology.org/

636d Multiscale Stochastic Analysis Of The Gal Switch In Saccharomyces Cerevisiae: Modeling And Experiments

Vinay Prasad, Dept. of Chemical Engineering, University of Delaware, 150 Academy Street, Newark, DE 19716 and K. V. Venkatesh, Chemical Engineering, Indian Institute of Technology Bombay, Dept of Chemical Eng, IIT Bombay Powai, Mumbai, 400076, India.

The GAL switch in saccharomyces cerevisiae is a prototypical eukaryotic genetic switch, and is well characterized at the biochemical and genetic levels. In this switch, galactose (an inducer) elicits a sigmoidal switch-like response for the expression of GAL genes. GAL genes are switched on by a transcriptional activator, Gal4p, which is under glucose repression. In a wild type strain, Gal4p itself is regulated by a repressor (Gal80p) and an inducer (Gal3p). It is noteworthy that both the regulators, Gal80p and Gal3p, are under the control of the GAL switch, constituting autoregulation. Since information is well documented for the GAL switch, numerous quantitative analyses have been carried out. Most of the analyses are carried out using models based on differential equations for the protein synthesis from genes under the GAL switch. However, the GAL genetic switch in a mutant yeast strain lacking Gal80p responds only to changes in Gal4p, which is in turn regulated by glucose concentration in the medium. The Gal4p synthesis is regulated by a kinase, Mig1p, whose unphosphorylated form enters the nucleus to repress the synthesis of Gal4p. The phosphorylation cycle is tightly controlled by glucose concentration in the cytoplasm. Thus, the levels of glucose determine the fraction of Mig1p available in the nucleus. In this contribution, we have modeled the mechanisms involved in the regulation of the GAL genes. We consider genes with one binding site for Gal4p (such as GAL3 and MEL1) and genes with two binding sites (such as GAL1 andGAL2). Also, Mig1p, the inhibitor controlled by glucose, has two binding sites for SUC2 and one binding site for GAL3 and GAL1. It is important to note that the GAL genes are regulated by Gal4p and also through Mig1p.

We seek to investigate the significance of two levels of control on gene expression. The other question we address is the relevance of inherent noise due to stochasticity on gene expression. In order to do this, we present results based on stochastic modeling using the stochastic simulation algorithm (SSA) [1] at the genetic scale. We combine the stochastic model with metabolic models through the chemical Langevin equation (CLE) approach, and with hybrid multiscale approaches using tau-leaping [2]. Finally, we present experimental results at the cellular level describing the variation in the phenotypic response, and characterize the effect of the stochastic variation at the genetic scale on the larger scales. This analysis is significant in that it provides a direct evaluation of the effect of intrinsic noise at different scales in saccharomyces cerevisiae, and on how it is modified by regulation.

References:

1. D.T. Gillespie, 'Exact stochastic simulation of coupled chemical reactions', J. Phys. Chem., 81(25), 2340 (1977).

2. A. Chatterjee, K. Mayawala, J. S. Edwards, and D. G. Vlachos, 'Time accelerated Monte Carlo simulations of biological networks using the binomial t-leap method', Bioinformatics, 21(9), 2136 (2005).

636e Rasgrp-Dependent Feedback Regulation Of Sos Contributes To Digital Erk Responses And Efficient Lymphocyte Activation

Jayajit Das[1], Jeroen Roose[2], Arthur Weiss[3], and Arup K. Chakraborty[1]. (1) Chemical Engineering, Massachusetts Institute of Technology, E19-238, 70 Massachusetts Avenue, Cambridge, MA 02139, (2) Department of Anatomy, School of Medicine, University of San Francisco at California, SanFrancisco, 94143, (3) Department of Medicine, The Rosalind Russell Medical Research Center for Arthritis and Howard Hughes Medical Institute, University of San Francisco at California, 94143

Recent work suggests that calcium fluxes can be stimulated by a few agonists, and this response is analog in nature. In contrast many other downstream responses require a larger threshold number of agonists (i.e., digital response). We have previously demonstrated that RasGRP-Sos crosstalk is the basis for optimal Ras-ERK activation in lymphocytes. Here we describe the results of synergistic computational and experimental studies which demonstrate that a positive feedback mechanism involved in the Sos-catalyzed activation of Ras results in a threshold for efficient downstream signaling during membrane proximal signaling events. Our results also show that clonal cells subjected to the same amount of stimulation partition in to two subpopulations (characterized by strong and weak ERK activation) because this feedback regulation leads to a bistable response. Our computational analysis predicts that a consequence of this positive feedback regulation is an emergent hysteretic behavior that allows stimulated T cells to sustain a high level of signaling even when the stimulus falls below the original activation threshold. This feature could confer stability of T cell activation to fluctuations in the stimulus or TCR levels. We are currently testing this with various experimental approaches.

636f Single-Cell Math Model With Experimental And Math Analysis Of The Apoptosis Regulatory Network

John M. Burke[1], John G. Albeck[2], Peter K. Sorger[2], and Douglas A. Lauffenburger[1]. (1) Biological Engineering, Massachusetts Institute of Technology, 77 Massachusetts Ave, 56-291, Cambridge, MA 02139, (2) Systems Biology, Harvard Medical School, 200 Longwood Ave, WAB 438, Boston, MA 02115

Apoptotic cell death is an essential physiological process misregulated in many diseases. A central challenge is to understand the quantitative features of apoptotic regulation, such as all-or-none activation of executioner caspases, and cell-to-cell variability in the time of death. Numerous theoretical models have been proposed to explain these behaviors; there have been few direct experimental tests of such hypotheses. We developed a math model of apoptotic regulation in parallel with extensive experimentation at the single cell level. Using live-cell reporters spanning the caspase activation network to monitor multiple signaling events simultaneously in individual cells, we identified the signal transduction steps in which the graded upstream signal induced by the initial death stimulus is converted into an all-or-none downstream response. Using model reduction and dynamical systems theory, control parameters are identified. Model predictions are verified numerically. Math results led to trials of protein overexpression, siRNA-mediated protein depletion, and pharmacological inhibition, and identified conditions under which all-or-none caspase activation fails. When caspase substrates are only partially cleaved, cells fail to choose decisively between death and survival, creating a potentially dangerous situation in which cells survive despite sustaining significant damage. Our computational and experimental studies have combined to generate a comprehensive model of the caspase regulatory network, which accurately predicts both normal and pathological behavior.

657a Deducing Hidden Transcription Networks And Regulatory Signals From Gene Expression Data

Mark P. Bryníldsen[1], Tung-Yun Wu[2], Shi-Shang Jang[2], and James C. Liao[1]. (1) Department of Chemical and Biomolecular Engineering, University of California, Los Angeles, 5513 Boelter Hall, 420 Westwood Plaza, Los Angeles, CA 90095, (2) Department of Chemical Engineering, National Tsing-Hua University, Hsinchu, 30043, Taiwan

Transcriptional regulation forms a bipartite network between genes and transcription regulators. Two key features of bipartite networks are the network architecture and regulatory signals. Such information is often implicitly imbedded in the network output, which in the case of transcriptional regulation would be the transcriptome. Here we develop a technique, Network Component Mapping (NCM), that deduces bipartite network connectivity and regulatory signals from data without the need for prior information. Network and regulatory signal inference in the absence of prior information is particularly attractive for situations when gene expression data is available from a poorly understood organism, or a poorly characterized environment. We demonstrate the utility of our approach by analyzing UV-vis spectra from mixtures of metabolites and gene expression data from Saccharomyces cerevisiae. From UV-vis spectra, hidden mixing networks and pure component spectra (sources) were deduced to a higher degree of resolution with our method than other current bipartite techniques. Analysis of S. cerevisiae gene expression from two separate environmental conditions (zinc and DTT treatment) yielded transcription networks consistent with ChIP-chip derived network connectivity. Due to the high degree of noise in DNA microarray data the transcription network for many genes could not be inferred. However, with comparatively clean expression data, our technique was able to deduce hidden transcription networks and instances of combinatorial regulation. For noisy data, NCM yields the sparsest network capable of explaining the data. In addition, partial knowledge of the network topology can be utilized by NCM as constraints.

657b A Novel Transcriptional Clustering Scheme Identifies Key Members Of The Eukaryotic Stress Response

Patrick D. McMullen, Department of Chemical and Biological Engineering, Northwestern University, Evanston, IL 60202, Luis A. N. Amaral, Chemical and Biological Engineering, Northwestern University, 2145 Sheridan Rd, TECH E136, Evanston, IL 60208, and Richard I. Morimoto, Biochemistry, Molecular Biology & Cell Biology, Northwestern University, IL.

All organisms are faced with challenges posed by their environments, so it is no surprise that biology has evolved a system to respond to environmental variations. This cellular stress response allows an organism to survive departures from its optimal environment, including thermal, radiative, osmotic, and chemical stresses. Deficiencies in cellular quality control have been linked to many neurodegenerative diseases, but the specific causes for these remain largely unknown.

We tap into the wealth of publicly available global transcriptional data to help answer these questions. Using S. cerevisiae as a model, we have implemented a new methodology for clustering genes based on transcriptional information that focuses on the global characteristics of the system. These genes can be further divided by the degree of connectedness each has to genes outside its cluster--a metric called the participation coefficient. We show that a gene's participation coefficient is linked to its cons ervation at the sequence level when compared to other model organisms. This reflects a novel way of identifying genes that play important roles in a biological system.

657c Comparison Of Multiple Stress Responses Of A Systematically Perturbed System

Bhaskar Dutta, University of Maryland, 2113 Chemical and Nuclear Engineering Building, College Park, MD 20742-2111, Harin H. Kanani, Chemical and Biomolecular Engineering, University of Maryland, College Park, 2113 Chemical and Nuclear Engineering Building, College Park, MD 20742-2111, John Quackenbush, Department of Biostatistics, Dana-Farber Cancer Institute, Mayer 232, 44 Binney Street, Boston, MA 02115, and Maria Klapa, Institute of Chemical Engineering and High-Temperature Chemical Processes, Foundation for Research and Technology-Hellas (FORTH), Stadiou St., Platani, Patras, Greece.

Perturbing a system and studying its response can provide insight about unknown systems. Biological systems, which are immensely complex, require them to be perturbed in multiple ways to elucidate the relationship among the system variables. Studying the response of systematically perturbed biological system in a dynamic fashion can reveal valuable clues about gene regulations, structure of regulatory networks and their robustness.

As a case study, *Arabidopsis thaliana* liquid cultures grown for 12 days under constant light and temperature were subjected to perturbations of (1) CO_2 level in their growth environment (2) Osmotic stress through addition of NaCl and (3) Trehalose (sugar) signal. Full genome cDNA microarrays (printed in TIGR) were used for dynamic gene expression profiling of the system subjected to the applied perturbations. According to the experimental design it was possible to study the dynamic effect of (a) large number of perturbations applied to the same system (b) combined perturbations and (c) comparison of the combined perturbation with the corresponding individual perturbations. Comparisons revealed individual perturbations are insufficient to predict not a good predictor of the combined response and the part of the regulatory networks that are conserved between the individual and combined responses depends on the system and stresses applied. Gene regulatory networks that are conserved over all the stress responses have biological implications in the context of metabolic pathways, chromosomal location and cellular compartmentalization.

This analysis has not only provided significant information about the physiology of plant in general and *A. thaliana* in particular, but has also generated a large experimental data set of a highly complex eukaryotic system under systematically applied multiple perturbations. As such it is expected to greatly contribute in enhancing the currently available experimental database and in advancing the systems biology computational and experimental toolbox.

657d Combining Computational And Experimental Methods To Identify Genes With Metabolic Functions

Jennifer L. Reed[1], Trina R. Patel[1], Keri H. Chen[1], Andrew R. Joyce[1], Margaret K. Applebee[2], Christopher D. Herring[1], Olivia T. Bui[1], Eric M. Knight[1], Stephen S. Fong[3], and Bernhard O. Palsson[4]. (1) Bioengineering, University of California, San Diego, La Jolla, CA 92093-0412, (2) Department of Chemistry and Biochemistry, University of California, San Diego, La Jolla, CA 92093-0412, (3) Department of Chemical and Life Science Engineering, Virginia Commonwealth University, Richmond, VA 23284-3028, (4) University of California, San Diego, 9500 Gilman Dr. Dept. 0412, La Jolla, CA 92093-0412

Genome-scale models of *Escherichia coli* K-12 MG1655 metabolism and regulation can predict growth phenotypes for wildtype and mutant strains in a variety of defined environments (1). With the development of high-throughput growth screens (such as Phenotype Microarrays, Biolog, Hayward, CA) the metabolic capabilities of organisms can be rapidly assessed (2). Such technology has been applied to study *E. coli* strains (2) and the results are used in this work. By evaluating each positive growth environment, in the context of a metabolic model, a subset of growth conditions can be found that can not be explained by the current metabolic model. These indicate that catabolic pathways or transporters are missing from the reconstructed metabolic network (3). A combined computational and experimental effort was used to uncover these missing metabolic components and the genes responsible for their activities. We use an optimization-based algorithm to identify potentially missing reactions (either enzymatic or transport) that are needed to enable the model to correctly predict growth, in agreement with experimental data. Knowing what enzymes or transporters to look for, potential candidate genes were identified through literature searches, sequence-based homology searches, context-based search methods, or through gene expression profiling. Subsequent experiments evaluating single gene knockout (4) phenotypes, gene expression measurements using quantitative RT-PCR, and enzyme assays (in one case), identified the genes responsible for missing metabolic enzymes and transporters. This combined computational and experimental effort led to updated functional assignments for eight open reading frames in *E. coli* with two new enzymatic activities and four transport functions (5). The study demonstrates the use of systems analysis to discover new metabolic and transport functions and their genetic basis by a combination of experimental and computational approaches.

REFERENCES: 1. Covert, M.W., Knight, E.M., Reed, J.L., Herrgard, M.J. & Palsson, B.O. (2004) *Nature* 429, 92-6. 2. Bochner, B.R., Gadzinski, P. & Panomitros, E. (2001) *Genome Res* 11, 1246-55. 3. Reed, J.L., Vo, T.D., Schilling, C.H. & Palsson, B.O. (2003) *Genome Biol* 4, R54. 4. Baba, T., Ara, T., Hasegawa, M., Takai, Y., Okumura, Y., Baba, M., Datsenko, K.A., Tomita, M., Wanner, B.L. & Mori, H. (2006) *Molecular Systems Biology*. 5. Reed, J.L, Patel, T.R., Chen, K.H., Joyce, A.R., Applebee, M. K., Herring, C.D., Bui, O.T., Knight, E.M., Fong, S.S., & Palsson, B.O. (2006) *PNAS* 103,17480-4.

657e Utilizing Variation In Gene Expression To Quantify DNA Damage

Michael G. Benton[1], Nathaniel R. Glasser[2], and Sean P. Palecek[2]. (1) Cain Department of Chemical Engineering, Louisiana State University, South Stadium Drive, Rm 110, Baton Rouge, LA 70803, (2) Chemical and Biological Engineering, University of Wisconsin - Madison, 1415 Engineering Drive, Madison, WI 53706

There is a high correlation between a chemical's ability to induce DNA damage and its ability to cause cancer. For this reason, it is crucial to rapidly identify substances which have the ability to damage DNA. We have adopted a genomics approach to better understand how the model organism Saccharomyces cerevisiae senses and responds to DNA damage, hoping that further elucidation of these complex pathways will lead to the design of improved methods for sensing and reporting the presence of a potential carcinogen. An ideal assay would not only identify when DNA damage is occurring, but report what type of damage is taking place (eg. single- or double-stranded damage) as well as quantifying the amount of damage that has occurred. In this work, photolithographic microarrays were used to quantify changes in the S. cerevisiae trancriptome in response to multiple types of DNA damage. This genome wide screen identified HUG1 as a gene that is highly induced by many types of DNA damage. This result led us to create a fusion in which the HUG1 promoter regulated expression of green fluorescent protein (GFP), creating a cell strain which produced detectable levels of fluorescence when DNA damage occurred. To yield a positive result (ie. conclude that DNA damage was occurring), GFP levels as measured by flow cytometry had to increase at least 2-fold over control cells not exposed to genotoxicity, with a p-value less than or equal to 0.05. These new cell lines allowed us to measure gene expression patterns in real-time in response to multiple cellular stresses, including genotoxicity. This ability enabled us to identify DNA damage response pathways that are differentially regulated based on the type of insult that has occurred, in a manner that is simpler and cheaper than traditional methods.

Biological Pathway-Centric Approach to Integrative Analysis of Array Data as Applied to Mefloquine Neurotoxicity

Abhishek Soni[1], Jerry Jenkins[1], Jonathan Hood[1], Michael Aschner[2], Chih-Thai Jiang[2], and Shankar Sundaram[1]
[1]CFD Research Corporation, Huntsville, AL 35805
[2]Vanderbilt University, Department of Pediatrics, Nashville, TN 37232

Abstract

Mefloquine, a potent anti-malarial therapeutic, is known to cause adverse neurological or psychiatric effects in a subset of patient population which has been reportedly linked with specific single nucleotide polymorphisms (SNPs). Development of a predictive clinical, molecular diagnostic screen capable for the *a priori* determination of the susceptibility of an individual to Mefloquine neurotoxicity is critical for safe administration. Towards achieving this goal, we present a novel systems-biology centered methodology, combining transcriptional profiling using primary rat cortical neurons exposed to Mefloquine and novel pathway identification/analysis tools enabling the development of a detailed intracellular pathway model to explain Mefloquine-induced neurotoxicity. Boolean methods are used to objectively down select potential SNP target candidates. The results from the proposed methodology have consistently identified PTK2B as the most likely SNP target across multiple data sets, and using multiple toxicity markers.

1. Introduction and Background

Rapid emergence of drug resistant infectious parasites represents an ever increasing global health burden. Malaria, caused by protozoa of the genus Plasmodia, is characterized as endemic in 101 countries and annually results in 300-500 million clinical cases with over 2 million deaths. Resistance to common anti-malarial drugs is reported globally, and is cited as the major contributing factor to the growing global malaria problem. Addressing this threat in 1970's, Walter-Reed Army Institute of Research (WRAIR) developed a 4-quinoline methanol compound named Mefloquine (trade name Lariam™), a potent, long lasting blood schizontocide, effective against drug-resistant Plasmodium falciparum [WHO, 2001]. Mefloquine (MFQ) is administered to the U.S. military service members stationed in areas known to harbor drug resistant malaria [NCID Bulletin, 2003], and is useful for both therapy (1250 mg/day) and chemoprophylaxis (250 mg/week). MFQ is able to maintain high concentrations in the blood for several weeks [Fidock, 2004], allowing for once weekly prophylactic administration to military forces, minimizing force reduction due to non-compliance [Dow, 2004]. However, MFQ is known to cause adverse neurological or psychiatric symptoms in as many as 25% percent of individuals taking MFQ at prophylactic doses and 70% of individuals taking MFQ at treatment doses. Adverse neurological effects include ataxia and mood changes [van Riemsdijk, 1997], thereby offsetting the positive prophylactic aspects of MFQ. Research efforts aimed at uncovering the primary neurotoxic mechanisms include extensive transcriptional profiling [Dow, 2003b; 2004; 2005], gap junction channel blockage [Cruikshank, 2004], acetylcholinesterase inhibition [McArdle, 2005], and adenosine-A2A receptor modulation [Weiss, 2003]. However, notwithstanding this extensive knowledge base, a fundamental mechanistic understanding of the sequence of events ultimately resulting in MFQ-induced neurotoxicity is currently lacking [Dow, 2003a].

Recent research strongly indicates that neurotoxicity associated with MFQ can be attributed to specific single nucleotide polymorphisms (SNPs) [Aarnoudse, 2006]. Identification of a critical set of predictive SNPs could form the basis of a clinical test to enable identification of at-risk individuals. Therefore, the objective of the current work is to develop and apply a novel systems biology-centered approach to construct a pathway model of cellular response to MFQ, and systematically identify a critical subset of species that could serve as the basis for a genetic screen. The current working hypothesis is that there exists a small subset of cellular components whose action is critical in determining the cellular response to MFQ. Small perturbations caused by genetic variation will alter the function of these critical species, and result in a change in the cellular response, ultimately manifesting itself as an altered observable physiological response. Rational (in contrast with trial-and-error) identification of optimal intracellular candidates for a robust genetic screen demands a fundamental understanding of cellular response to MFQ.

In order to achieve this goal, we first quantify the transcriptional response of primary neurons exposed to MFQ, and using the microarray data construct an intracellular pathway model of cellular response to MFQ based on information from relevant databases, scientific literature, and expert guidance. Next a prioritized list of cellular processes associated with MFQ exposure is generated using novel computational analysis methods. The final step is the verification and validation of the intracellular pathway model using gene suppression/expression techniques in order to quantify the impact of genes identified as critical to cellular response to MFQ.

2. *Methods*

This paper presents the application of a systems biology-centered methodology, combining experimental transcriptional profiling and computational analysis, for the identification of cellular response mechanisms to MFQ exposure. The research methodology to achieve this objective is outlined in the subsequent sections.

2.1 Experimental Protocol

Materials: Mefloquine hydrochloride (MFQ) was purchased from Sigma (St. Louis, MO). All other chemical were also purchased from Sigma unless otherwise stated. Coverslips for cell culture were purchased from Carolina Biological Supply (Burlington, NC). All tissue culture media and supplements were purchased from Invitrogen (Carlsbad, CA), except for Hyclone Fetal Bovine Serum and Hyclone F12, from VWR International (Suwanee, GA).

Cell Culture Conditions and Mefloquine Exposure: Rat cortical neuron cultures were prepared from E17 rat pups, as previously described [McLaughlin, 1998]. The isolated cortices were then transferred to a petri-dish containing 0.6% (w/v) trypsin in cold Hank's balanced salt solution (HBSS) for 30 minutes. After two washes in HBSS, the cortical tissues were mechanically dissociated with a glass pipette. Dissociated cortical cells were plated on poly-L-ornithine-treated glass coverslips in 6-well plates, using a plating medium of glutamine-free DMEM-Eagle's salts (Invitrogen, Carlsbad, CA), supplemented with Ham's F12 (Hyclone, Logan, UT), heat-inactivated fetal bovine serum (Hyclone, Logan, UT), and penicillin/streptomycin (Sigma, St. Louis, MO), at a density of 700,000 cells per well. After two days in vitro, non-neuronal cell division was halted by a one day exposure to 10 μM cytosine arabinoside (Sigma, St. Louis, MO), and cultures were shifted to neurobasal media (Invitrogen, Carlsbad, CA), supplemented with B27 (Invitrogen, Carlsbad, CA) and penicillin/streptomycin. Cells were maintained by changing the media every 2-3 days and grown at 37 °C in a humidified atmosphere of 5% CO_2 in air. Cells were treated three weeks after isolation with MFQ, prepared from a 100 mM stock solution in dissolved DMSO with treatment buffer, for 24 hours, at 37 °C in a humidified atmosphere of 5% CO_2 in air. Treatment buffer consisted of MEM (Invitrogen, Carlsbad, CA) supplemented with HEPES, 2x N2 media supplement (Invitrogen, Carlsbad, CA).

Primary rat cortical neuron viability, was determined after a 24 hours exposure with LIVE/DEAD Viability/Cytotoxicity Kit (Molecular Probes, Eugene, OR). Internal consistency of the data was determined using both the MTT (Sigma, St. Louis, MO), and LDH (Sigma, St. Louis, MO) assays. Optimal MFQ concentration for microarray experiments was also obtained using viability/cytotoxicity assays which indicate that MFQ results in decreased primary rat cortical neuron viability following a 24 hour exposure. From the dose-response data (results not shown), the LD_{50} of MFQ in primary rat cortical neurons is calculated to be 8.9 μM, which is comparable to previously measured values [Dow, 2003a; 2003b; 2005].

Microarray Data Sets: To extract RNA from the neuron cultures, neurons were harvested by single-step acid guanidinium thiocyanate-phenol-chloroform extraction [Chomczynski and Sacchi 1987; Chomczynski 1993; Chomczynski and Mackey 1995]. The RNA was sent to VMSR and microarray measurements were commenced. All hybridization experiments were carried out using the Applied Biosystems Rat Genome Survey Array, which contains approximately 27,000 gene probes, and all measurements were conducted using the AB 1700 Chemiluminescence Microarray Analyzer. Two additional existing data sets [Dow, 2003b; 2005] were included in the analysis in order to ascertain the relative variation in cellular response with different primary cell types. Both data sets were produced using primary neuronal cultures of fetal rat forebrain exposed to 80 μM concentrations of MFQ for 5 minutes, then the media was then replaced with fresh minimal essential medium and the neurons were incubated for 24 hours. Microarray measurements were performed after the 24 hour period using the Affymetrix Rat Genome U34A Chip (approximately 8,800 gene probes).

2.2 Construction of Intracellular Pathway Model

The first step in constructing a computational model of the cellular response to MFQ stimulation is the construction of a relevant pathway model. An overview of the systematic process that uses microarray data to generate a computation-ready pathway model file is outlined in Figure 1 and summarized in the subsequent sections.

2.2.1 Identification of Significant Genes: The first step in the creation of a pathway model is to identify the subset of differentially expressed genes of biological significance, and associate them with their respective gene products. Normalized microarray data of primary forebrain neuronal response to MFQ exposure [Dow, 2005] was formally obtained from Dr. Geoffrey Dow (Walter Reed Army Institute of Research, WRAIR). Affymetrix Rat Genome U34A array platform specific information was retrieved from the Gene Expression Omnibus (GEO) database and used to associate the individual Affymetrix Probe Set ID with the respective Entrez Gene ID. This was performed in an automated fashion using an in-house software module developed for GEO database query and retrieval. Fold change was computed using the mean expression values for each gene, divided by the DMSO control value. Significance analysis of microarrays (SAM) [Tusher, 2001] was performed using the MultiExperiment Viewer (MeV) v4.0 (TM4 Development Group) to identify differentially regulated genes using 792 column perturbations and a false discovery rate (FDR) of 5%. SAM performed on the 80 μM MFQ data set [Dow, 2005] yielded 640 significant genes. Microarray data (AB1700 Rat Genome Survey Array) received from VMSR was analyzed using fold change exclusion method (2-fold) [Knudsen, 2004], as the microarray data contained two replicates for the 3 hour and 12 hour exposure data. Genes that were found to be 2-fold upregulated or down regulated in both replicates were included. Fold change exclusion of the 3 hour MFQ exposed replicates yielded 594 significant genes and the 12 hour MFQ exposed replicates yielded 1859 significant genes. The list of significant genes was used to identify significant functional categories and canonical pathways using an ontology analysis.

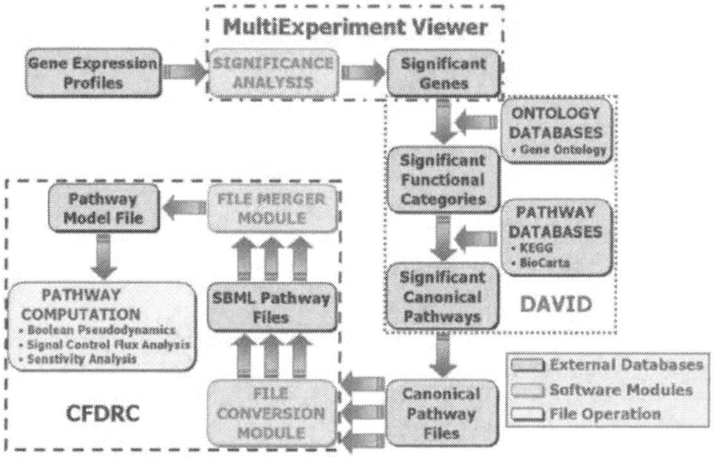

Figure 1: Overview of the pathway generation process

2.2.2 Canonical Database Selection: Association of genes with specific biochemical pathways requires an understanding of the biological context of a specific gene within a pathway model. Many databases exist that contain qualitative or quantitative descriptions of protein and signaling interactions (e.g. GeneNet [Kolpakov, 1998], BIND [Bader, 2001], KEGG [Kanehisa, 2002]) and enzyme kinetics (e.g. BRENDA [Schomburg, 2002]). In this task we used the Kyoto Encyclopedia of Genes and Genomes (KEGG) database as our source of canonical pathway files (http://www.genome.jp/kegg). The Kyoto Encyclopedia of Genes and Genomes (KEGG) database was used to establish relationships between genes obtained from transcriptional microarray data and their respective protein products and to obtain canonical metabolic and signaling pathways that form the basis of MFQ neurotoxicity pathway model development. KEGG pathway files contain all of the information associated with a given canonical pathway, and are maintained in KEGG Markup Language (KGML) file format.

2.2.3 Canonical Pathway Construction: Entrez IDs of significant genes were used to identify statistically relevant KEGG canonical pathways using the Database for Annotation, Visualization and Integrated Discovery (DAVID) [Dennis, 2003]. DAVID incorporates ontological (e.g. GO) information and KEGG canonical pathway databases to simultaneously identify and visualize statistically significant functional categories and canonical pathways from a list of biologically relevant genes. The relevant canonical pathways for the 80 μM (5 minute exposure) data [Dow, 2005], and the 10 μM data exposed for 3 and 12 hours are shown in Figure 2. The results of the DAVID analysis produced canonical pathways primarily involved in cellular communication, metabolism, apoptosis, and Ca^{2+} signaling for all three conditions considered. Pathways that are active in multiple conditions are apoptosis, Ca^{2+} signaling, JAK-STAT signaling, neuroactive ligand-receptor pathways, and cytokine-cytokine receptor interaction. Only one canonical pathway was observed in all three, the neuroactive ligand-receptor pathway, containing a variety of ligand-receptor interactions such as dopamine, acetylcholine, and bradykinin. Significant canonical pathways identified using DAVID were used to construct a biochemical pathway model of cellular exposure to MFQ.

2.2.4 Creation of Pathway Model File: Unfortunately, the KGML format is not an accepted standard format for computational work in systems biology. Therefore, the KGML files were converted to the portable standard Systems Biology Markup Language (SBML) [Hucka, 2003; Hedley, 2001] format. An in-house developed python-based software toolset was developed for the automated conversion of KGML files to the SBML format. Additonal capabilities of this toolset include the ability to query and retrieve the necessary ontological information from the KEGG ontology database. The KEGG ontology database links genomic information to proteomic representation through an expanded ontology [Ogata, 1999; Kanehisa and Goto, 2000; Kanehisa, 2006]. KEGG is easily accessed using the Simple Object Access Protocol (SOAP) library for Python, SOAPpy, enabling automation of the retrieval of information from the KEGG GENE, KEGG ENZYME, KEGG COMPOUND and KEGG ONTOLOGY databases, necessary to establish relationships between microarray data and resultant protein products. KGML files were converted to the SBML file format by mapping the specific reaction and species types onto the SBML standard. Once generated, the individual files are merged into a single SBML file describing the cellular response of the pathway model to MFQ. The resultant pathway model was then autmented using external information and used for further analysis to identify potential SNP targets.

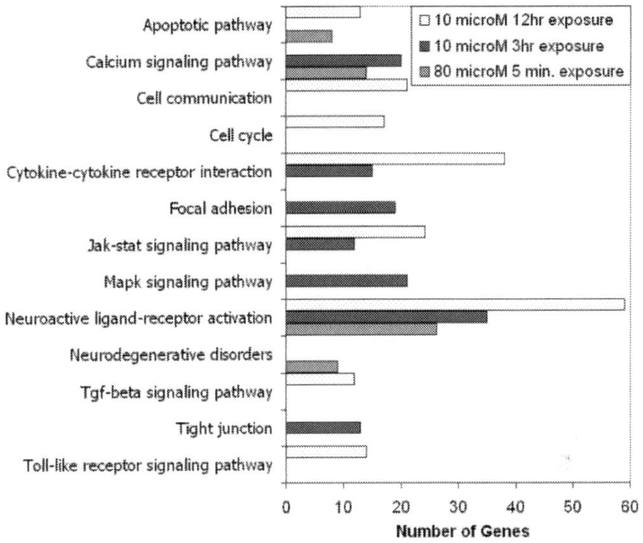

Figure 2: KEGG canonical pathways identified by DAVID for the three data sets considered.

2.2.5 Addition of Mefloquine and ABCB1 to the Pathway Model: Construction of a pathway model of cellular interaction requires that we add all relevant information to the model to enable a more complete analysis. An important piece of information to include in the model is the physiological mechanisms responsible for allowing MFQ to cross the blood-brain barrier, and concentrate in neural tissue. The multiple drug resistance (MDR1) P-glycoprotein (ABCB1), an ATP-binding cassette (ABC) transporter, is an ATP-dependent efflux protein that is expressed in the capillary endothelial cells that line the blood brain barrier. MFQ has been shown to interact with the substrate-binding site of human ABCB1 and to inhibit the transport activity of ABCB1 [Riffkin, 1996]. The inhibition of ABCB1 by MFQ has been shown to act to inhibit MFQ efflux resulting in the concentration of MFQ in neural tissue, exacerbating the neurotoxic effects. Single-nucleotide polymorphisms (SNPs) of MDR1 have been shown to be associated with decreased ABCB1 expression and function [Hoffmeyer, 2000]. Specifically, ABCB1 1236TT, 2677TT and 3435TT genotypes have been associated with neurotoxic effects of MFQ exposure [Aarnoudse, 2006].

Recent literature also supports the view that MFQ acts to destabilize Ca^{2+} homeostasis, resulting in endoplasmic reticulum (ER) stress [Dow 2003b; 2005], ultimately leading to the up-regulation of DNA damage-inducible transcript 3 (DDIT3/GADD153), an ER stress response gene, and the activation of caspases. The primary challenge in adding MFQ to the pathway model lies in the fact that MFQ does not have known specific cellular target(s). Given the overwhelming evidence of destabilized Ca^{2+} homeostasis, two targets of MFQ were selected. CACNA1 (voltage dependent Ca^{2+} channel) and GNA14 (guanine nucleotide binding protein that activates PLCβ were chosen as entry points into the pathway model for MFQ [Lim and Go, 1985]. Both CACNA1 and GNA14 act to rapidly increase intracellular Ca^{2+} concentrations and destabilize homeostasis. The primary assumption we are making is that when MFQ is present CACNA1 and GNA14 are active, and when MFQ is absent they are inactive. The addition of nodes based on physiological information illustrates a major benefit of the current framework: enhanced flexibility through the incorporation of known experimental information, and/or insight provided by scientific literature.

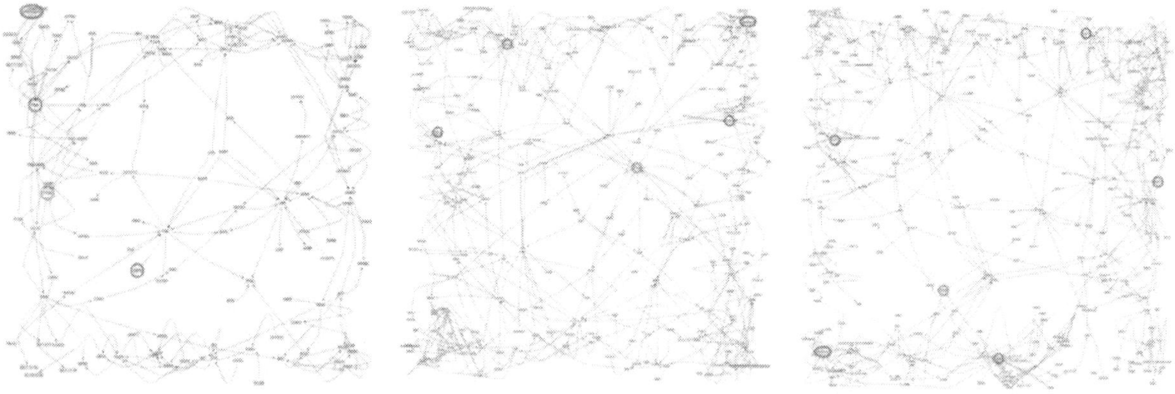

(a) (b) (c)

Figure 3: Final pathway model files for (a) primary rat forebrain cell response to a 5 minute exposure of 80 μM MFQ [Dow, 2005], and primary rat cortical neuron response to (b) 3 hour or (c) 12 hour exposure to 10 μM MFQ. MFQ (red), Ca^{2+} (blue), PTK2B (orange), DDIT3 (green), CASP12 (green), and GADD45 (green). Pathway (a) contains 127 species and 177 reactions, pathway (b) contains 224 species and 306 reactions, and finally pathway (c) contains 272 species and 353 reactions.

2.2.6 Endpoint Selection: Selection of outputs of the biochemical pathway model was based upon the current biological understanding of cellular response to MFQ exposure and the transcriptional profile of cellular response to MFQ exposure. Assuming that MFQ acts to increase intracellular Ca^{2+} concentrations resulting in ER stress [Dow 2003b; 2005], the ER stress leads to increased expression of unfolded protein response (UPR) genes such as DNA damage-inducible transcript 3 (DDIT3). Prolonged ER stress leads to increased expression of UPR and apoptotic genes such as caspases. Therefore, DDIT3 was selected as an output for the pathway model describing 3 hour exposure to 10 μM MFQ. Additionally, DDIT3 and caspase-12 were selected as outputs for the biochemical model describing 10 μM MFQ exposure for the 12 hour dataset. Based on the selected input and output nodes, pathway models were constructed for each of the conditions (Figure 3). Quantitative analysis of these pathway models to identify potential SNP targets is discussed in the subsequent sections.

2.3 *Objective ranking of sub-cellular processes using pathway analysis*

Having obtained a pathway model file, computational analysis can be carried out to objectively rank order all sub-cellular processes based on their contribution to MFQ-induced neurotoxic response. A novel computational methodology has been developed that utilizes the Boolean pseudo-dynamic approach coupled with a modified signal control flux technique to obtain a ranking of the sub-cellular mechanisms. A detailed description of the algorithmic framework is beyond the scope of the current paper;

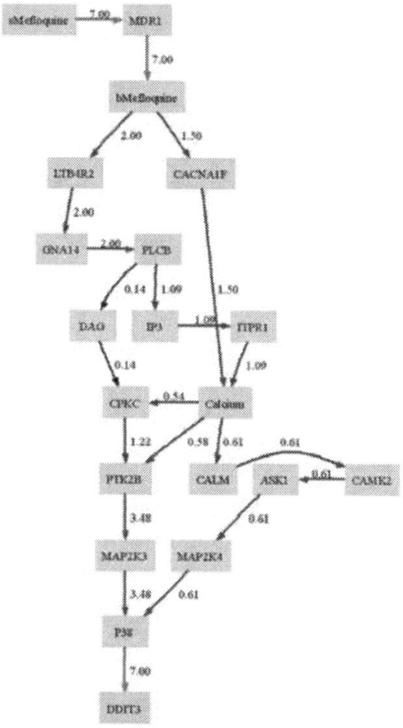

Figure 4: CEF values for the pathway from MFQ to DDIT3 for the 3 and 12 hr pathway model.

however relevant details can be found elsewhere [Soni, 2007, Soni, 2007a]. The relevant algorithmic features are discussed briefly. In order to obtain the pseudo-dynamic state trajectories the SBML biochemical pathway model is first parsed to obtain species logical rules, containing Boolean operators. The form of the logical rule is dictated by the reaction type and is focused around their respective species. Specifically, species that can be activated by multiple species form an 'OR' rule, inhibition reactions form an 'AND NOT' rule, and binding/association reactions form an 'AND' rule. Next, a Boolean pseudo-dynamic (BPD) state update is carried out to yield the state trajectories [Albert, 2003; Chaves, 2005; 2006, Li, 2006]. In order to accurately account for all possible initial values for network nodes, a random distribution of the intial states is used and an ensemble of simulations is employed to best characterize the system. The output from BPD update is an array of 0's and 1's describing the state trajectories in the system, where 0 represents an inactive node, whereas 1 represents an active node at that instant. This state information is then used to determine the set of active signal transfer routes from the input to the output, and subsequently the control effective flux (CEF) for a reaction in the active pathway is computed. The CEF value is the measure of importance of a reaction in maintaining the overall network phenotype [Stelling, 2002]. Therefore, a reaction with a high CEF value is more important in maintaining the cellular function and is thereby a critical interaction.

3 Results

In order to characterize the cellular response to MFQ, an ensemble of 300 simulations were performed utilizing DDIT3 as the logical output for the 3 hr biochemical pathway model and DDIT3 and caspase-12 as the logical output for the 12 hr biochemical pathway model. When comparing DDIT3, a minor difference was observed in the CEF values of the 3 hr simulation and 12 hr simulation, which, on average, was only 0.5% higher than that of the 3 hr simulation. Seven unique paths were identified from the logical input (i.e. serum MFQ) and the selected logical output (Figure 4). Each reaction in the seven active paths is represented based upon its calculated CEF value, whereby large CEF values (red) indicate a more important interaction relative to all other interactions (blue or black). The CEF values are considerably high in the absorption of serum MFQ into the brain through the MDR1 P-glycoprotein. This is an expected result, as MDR1 represents the only mode of MFQ entry into the brain. Once MFQ enters the neuron, it acts to increase intracellular Ca^{2+} concentrations, which may be achieved via activation of Ca^{2+} channels (i.e. CACNA1F) and/or activation of G-protein coupled receptors (i.e. GNA14), such as acetylcholine receptors. CEF values indicate that reactions downstream of PTK2B are important relative to other interactions (i.e. there is a preference for MAP2K3), whereby the value is greater than 5X the nearest value. PTK2B encodes for a non-receptor protein tyrosine kinase kinase-2β (Pyk2), a member of the focal adhesion kinase family (FAK), involved in Ca^{2+}-induced regulation of ion channels and activation of the MAP kinase signaling pathway [Avraham, 2000]. Pyk2 undergoes tyrosine phosphorylation and activation in response to increases in intracellular Ca^{2+} concentration or protein kinase C (PKC) activation. Active Pyk2 phosphorylates mitogen-activated protein kinase 3 (MAP2K3 or MKK3), which phosphorylates p38 [Pandey, 1999], ultimately leading to the activation of DDIT3. The results from the analysis of the 3 hr and 12 hr biochemical pathway models suggest that the mechanism of MFQ induced stress response is conserved and stable over the duration of exposure.

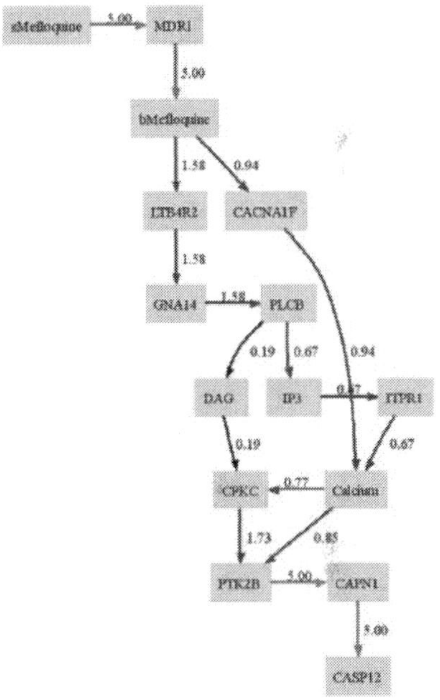

Figure 5: CEF values for the pathway from MFQ to caspase-12 for the 12 hr pathway model.

Similarly, CEF calculations were performed to analyze the information flux distribution for cellular response to MFQ leading to the activation of caspase-12, which was determined to be significant in the 12 hr biochemical pathway model. The five unique paths leading from serum MFQ to caspase-12 are illustrated in Figure 5. As expected, the CEF values are high in the absorption of serum MFQ into the brain via the MDR1 P-glycoprotein. CEF values indicate that reactions involving and downstream of PKC are important relative to other interactions. This includes PTK2B modulation of calpain, which activates caspase-12 via proteolytic cleavage. Due to differences between the path to DDIT3 and caspase-12, the CEF values are altered, as they reflect the change in network topology.

3.1 SNP Target Selection

The CEF analysis indicated that the PTK2B-mediated stress response is conserved and stable over the duration of exposure. PTK2B was selected, as its activity is specific to endoplasmic reticulum stress and the disruption of Ca^{2+} homeostasis. The PTK2B gene encodes for a non-receptor protein tyrosine kinase kinase-2β (Pyk2), a member of the focal adhesion kinase family (FAK), involved in Ca^{2+}-induced regulation of ion channels and activation of the MAP kinase signaling pathway [Avraham, 2000]. By modulating ion channel function and MAPK signaling, Pyk2 can have both acute and chronic effects on the central nervous system. Pyk2 is also highly expressed in the central nervous system (Figure 6), specifically the hippocampus, which is coincidentally the location of highest MFQ concentration [Baudry, 1997]. The primary structure of Pyk2 is a sequence of 1,009 amino acids, but to date no secondary structure has been determined. Pyk2 has a large number of known single nucleotide polymorphisms (SNPs) listed on ENTREZ SNP database Therefore, PTK2B is an ideal selection for gene silencing based upon the CEF values, and it's specificity to Ca^{2+} and related stresses. Experimental testing of PTK2B as a target is currently underway using siRNA technology, whereby PTK2B is silenced in rat primary cortical neurons.

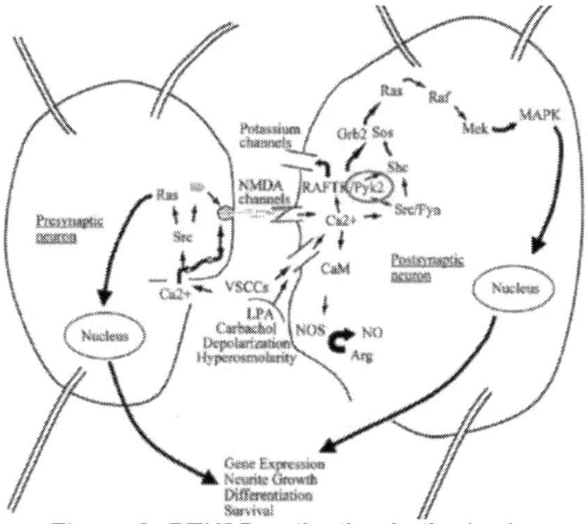

Figure 6: PTK2B activation in the brain.

4 Summary

In this work, the relevant canonical pathways involved in describing cellular response to MFQ exposure were identified using the microarray data, and tools for the automate creation of the full pathway model describing cellular response to MFQ were developed. A novel computational algorithm based on the Boolean pseudo-dynamice framework was developed to analyze the pathway model and obtain an objective ranking of the critical interactions, that could be potential targets for SNP analysis. The algorithmic framework was found to be robust and analysis of the experimental data identified PTK2B as a potential SNP target. The proposed systems biology-centered methodology has the capability to identify potential intracellular targets, ultimately providing a rational basis for the development of a genetic screen to identify individuals predisposed to Mefloquine neurotoxicity.

References

Aarnoudse et al. (2006) *Clin Pharmacol Ther.* **80**:367
Albert, R. (2005) *J Cell Sci.* **118**:4947
Avraham et al. (2000) *Cell Signal.* **12**:123
Bader et al. (2001) *Nucleic Acids Res.* 29:242-5.
Baudry et al. (1997) *Pharm.Pharmacol. 49*:1086-1090.
Chaves, M. et al. (2005) *J. Theor. Biol.* **235**:431
Chaves, M. et al. (2006) *IEE Proceedings in Systems Biology* **153**:154
Chomczynski and Mackey. (1995) *Biotechniques* **19**:942
Chomczynski and Sacchi. (1987) *Anal. Biochem.* **162**:156.
Chomczynski. (1993) *Biotechniques* **15**:532
Cruikshank et al. (2004) *Proc Natl Acad Sci USA.* **101**:12364-12369.
Dennis et al (2003) *Genome Biol.* **4**:R60
Dow. (2003a) *Malar J.* **2**:4
Dow et al. (2003b) *Malar J.* **2**:14
Dow et al. (2004) *Antimicrob Agents Chemother.* **48**:2624

Dow et al. (2005) *Genomics*. **86**:539
Fidock et al. *Nature Reviews Drug Discovery*. **3**:509-520. (2004)
Hedley et al. (2001) *Phil Trans Roy Sco London A*. **359**:1073
Hoffmeyer et al. (2000) *Proc Natl Acad Sci USA*. **97**:3473
Hucka et al. (2003) *Bioinformatics*. **19**:524
Kanehisa and Goto (2000) *Nucleic Acids Res*. **28**:27
Kanehisa et al. (2002) *Nucleic Acids Res*. **30**:42.
Kanehisa et al. (2006) *Nucleic Acids Res*. **34**:D354.
Knudsen, S. (2004) Guide to analysis of DNA microarray data, 2nd ed, John Wiley & Sons, Inc.
Kolpakov et al. (1998) *Bioinformatics*. **14**:529
Li, S. et al. (2006) *PLoS Biology* **4**:e312
Lim and Go (1985) *Clin Exp Pharmacol Physiol*. **12**:527
McArdle et al. (2005) *Neuropharmacology*. **49**:1132
McLaughlin et al. (1998) *Neuroscience*. **86**:279.
National Center for Infectious Diseases. (2003) Prescription drugs for prevention of malaria. [Online] Accessed April 2006.
Ogata et al. (1999) *Nucleic Acids Res*. **27**:29
Pandey et al. (1999) *J Biol Chem*. **274**:10140.
Riffkin et al. (1996) *Biochem Pharmacol* **52**:1545
Schomburg et al. (2002) *Nucleic Acids Res*. **30**:47
Soni, A. et al., (2007) *IET Sys Bio*., 2007, (under review).
Soni, A., et al. (2007), AIChE (http://aiche.confex.com/aiche/2007/techprogram/P96373.HTM)
Stelling, J. et al. (2002) *Nature*. **420**:190
Tusher et al. (2001) *Proc Natl Acad Sci USA*. **98**:5116
van Riemsdijk et al. (1997) *Eur J Clin Pharmacol*. **52**:1
Weiss et al. (2003) *Neurology*. **61**:S101.
World Health Organization. (2001) The use of antimalarial drugs. Report [Online] Accessed April 2006.

657g Knowledge Driven Matrix Factorization To Reconstruct Modular Gene Regulatory Network

Zheng Li[1], Shireesh Srivastava[2], Yang Zhou[3], Rong Jin[3], and Christina Chan[4]. (1) Department of Chemical Engineering and Material Science, Michigan State University, 2527 Engineering Building, East Lansing, MI 48824, (2) Michigan State University, 2527 Engineering Building, East Lansing, MI 48824, (3) Computer Science and Engineering, Michigan State University, East Lansing, MI 48824, (4) Chemical Engineering and Material Science, Michigan State University, 2527 Engineering Building, East Lansing, MI 48824

Reconstructing gene regulatory networks from micro-array data is important for understanding the underlying mechanism behind cellular processes. Numerous studies have been devoted to addressing this problem. One of the more popular methods is to view the gene regulatory network as a Bayesian inference network, and to apply structure learning methods to determine the topology of the gene regulatory network. There are, however, several major defects with the Bayesian structure learning approach for reconstructing gene regulatory networks such as the computational cost, non-unique solutions, does not take into account the special topology of gene regulatory network, and inefficiency in exploiting prior knowledge.

In this paper, we present a knowledge driven matrix factorization framework for reconstructing modular gene regulatory network. Unlike the Bayesian network approach, this approach views the reconstruction of gene regulatory network as a matrix factorization problem. The key advantage of the proposed framework in comparison to a Bayesian approach is its computational efficiency and power in exploiting the domain knowledge. Gene expression data is initially used to estimate the correlation matrix W. A matrix factorization algorithm is developed to decompose W into W=MCM' with a priori knowledge of gene ontology integrated to regularize M and C matrices. Here, matrix M provides the information about the gene regulatory modules. Each element of M indicates the confidence of assigning a gene to a module. Matrix C reveals the regulatory interaction between modules. Each element of C indicates the likelihood of two regulatory modules to interact with each other.

The matrix factorization approach was initially applied to yeast cell cycle expression data and the resulting modules were compared to that of the Bayesian approach, e.g. using the gene module network in Genomica [1]. Upon comparing the precision and recall, our approach performed better than the Bayesian method. We then applied the matrix factorization approach to gene expression from HepG2 cells exposed to different free fatty acids and identified biologically meaningful modules [2]. For example, all the genes in the TCA cycle were found in one module. Modules containing genes relevant to electron transport chain complex I, electron transport chain complex IV, ATP synthesis, were similarly identified. Analyzing the C matrix identified the connection or interaction between the modules. We found that many relevant modules predicted by this approach were indeed biologically related, for example the TCA module was found to have a strong interaction with the complex I, complex IV, and ATP modules. Indeed, it is well-known that these four processes are connected and involved in energy production. In summary, we have developed a knowledge driven matrix factorization approach that can simultaneously identify gene modules and how these modules are connected in a network.

1. Segal E. et al. (2004) A module map showing conditional activity of expression modules in caner, Nature Genetics, 36 (10):1090-1098. 2. Srivastava, S.; Chan, C. (2007) “Hydrogen peroxide and hydroxyl radical mediate palmitate-induced cytotoxicity to hepatoma cells: relation to MPT", Free Radical Research, Free Radical Research, 41(1),38-49.

AUTHOR INDEX

AUTHOR INDEX

American Institute of Chemical Engineers
3 Park Avenue
New York, New York 10016-5991

ISBN: 978-1-60423-838-9